More Than Just Roots: Unveiling the Layers of DNA Ancestry Tests

Sachin

TABLE OF CONTENTS

Page

INTRODUCTION

DNA ancestry tests have become a significant cultural object within the last decade. DNA testing company advertisements represent DNA tests as a fun gift idea to give to friends, family, or yourself to learn more about the family story that is presumed to be hidden inside one's DNA. The concept of "ancestral DNA" imagines one's DNA as not just the genetic material passed down from one's parents, but also a combination of various smaller pieces of DNA that have been passed down by the various ancestors in one's family tree. DNA testing companies claim they can use their DNA tests to break apart a person's DNA and discover the regional origins of the various pieces of that person's DNA. Through this analysis, they argue, DNA tests can quantify in percentages the various regional origins, which these companies often call "reference populations," that make up one's DNA (i.e., 55% British, 25% West African, 20% Italian). Consumers are instructed to understand these test results as an objective source of knowledge about who they are and where their ancestors came from.

Commercial genetic ancestry testing began to gain steam in the early 2000s.[1] Genetic ancestry testing reached its peak in 2017 and 2018, when companies such as Ancestry and 23andMe saw exponential growth in their consumer base. In 2017, the number of people who had taken a DNA test more than doubled and industry estimates suggest that by the end of the year the total number of people who have taken a DNA test since 2013 exceeded 12 million.[2] In the same year, Ancestry reported that they had tested more than seven million people, with two million tests being sold in the last four months of the year, and 23andMe reported selling more than 3 million tests by the end of 2017.[3] The third and fourth largest sellers following behind Ancestry and 23andMe were MyHeritage DNA and FamilyTreeDNA, respectively. By the end of 2019, when there was a decline in sales, an estimated 26 million Americans had bought a DNA ancestry test, with most of these consumers buying tests from the two largest players in the U.S. market, Ancestry and 23andMe.[4] This increase in sales is correlated with a significant increase in advertisement spending in 2016 and 2017. Ancestry spent $109 million in advertisements in 2016 while 23andMe spent $21 million in ads.[5]

DNA ancestry test results, or one's presumed DNA makeup, are often invoked when someone makes a claim of belonging to certain racial, ethnic, and national groups and when someone makes a claim about the (in)accuracy of one's family story. The prominence of DNA testing raises questions regarding how these tests have become integrated into how we talk about race and ethnicity in the U.S. and how we understand our connection to people in other parts of the world.

A moment that illustrates this practice, and inspired this book, occurred when I sat in on an undergraduate class as a graduate student. The students were asked if they had ever taken a DNA test. To my surprise, more than half of the students had. When discussing their DNA test results, many of the students mentioned that the test results often contradicted how they

---

[5] Regalado, "2017 was the year."

understood their ancestry or how their family talked about their family history. They interpreted these results as proving that their family was wrong or misunderstood their family history or even that their family had lied about their ancestry. One student specifically mentioned how a relative had said that they had indigenous ancestry for years, although when that same relative took a DNA test their results did not show any percentage of indigenous ancestry. The student interpreted this result to mean that their relative had been "lying" about their indigenous ancestry. However, as the conversation progressed, the student revealed that this relative had a tribal identification card. The student's relative having a tribal ID means that they are recognized as a member of an indigenous nation, regardless of the results of a DNA test. It stood out to me at that moment that the student took a DNA test result as a more accurate source of knowledge about their relative's ancestral past in comparison to the relative's identification card. This moment offered an example of how DNA tests are shaping how people understand and discuss race and ethnicity in the United States.

Since most of us do not have direct experience with DNA, many are gaining knowledge about DNA, how DNA works, and the connections between DNA, ancestry, race, ethnicity, and nationality through the media, especially from advertisements created by DNA ancestry testing companies. These ads have an impact on how we understand what DNA is and the relevance it has to our lives. In this book, I analyze the television advertisements created by the DNA ancestry testing companies Ancestry, 23andMe, and MyHeritage DNA and explore how these ads represent the connections between genetics, ancestry, race, ethnicity, and nationality. I also explore how these ads represent ancestry travel, or travel that is based on the results of one's ancestry test, and how communities and geographical regions across the world are represented in these ads. Lastly, I consider how these ads represent what racism is and the root causes of racism and how they propose that DNA tests can serve as tools to challenge racism.

Research Questions

For this book, I chose to analyze Ancestry, 23andMe, and MyHertiage advertisements because they are the three top DNA ancestry testing companies in the U.S. The research questions that guided my analysis of these advertisements include: 1) How are the histories of racial science and biological essentialism shaping both the DNA ancestry tests themselves and the claims being made by DNA ancestry testing companies in their marketing? 2) What are the connections made between genetics, race, ethnicity, ancestry, and nationality in these ads? 3) How is travel based on the results of one's DNA test represented in these ads? 4) How are populations and regions outside the U.S. represented in the ads that focus on the theme of ancestry travel? 5) How is racism and the root causes of racism defined in these ads? And 6) What are some of the social and political consequences that uncritical acceptance of these DNA tests can have on communities of color and on efforts to fight against racism?

Theoretical Framework

Race and Racism

My understanding of race and racism has been informed by the work of Howard Omi and Michael Winant (2015) and Ronald Mize (2013). Omi and Winant (2015)'s theory of racial formation is instructive for thinking through how the meanings associated with racial categories and racially stratified social structures have changed over time. In *Racial Formation in the United States*, Omi and Winant propose a new way of looking at the concept of race that understands race to be a "unstable and 'decentered' complex of social meanings constantly being transformed by political struggle."[6] They argue that while race is often assumed to be grounded in ideas about human biological diversity it is actually a socially constructed concept that takes for granted the idea that human beings can be grouped together into distinct groups based on perceived differences in physical characteristics, including skin color, hair texture, and facial structure. Perceived human physical differences are associated with more profound differences in human characteristics, such as intelligence, temperament, and athletic ability, and are used to "justify or reinforce social differentiation."[7] Omi and Winant ground race in the social and political, seeing it as a concept used to justify unequal treatment and distribution of social, political, and economic resources.

From this understanding of race, Omi and Winant develop their theory of racial formation. Racial formation is defined as "the sociohistorical process by which racial identities are created, lived out, transformed, and destroyed."[8] The two key components of this process are racialization and racial projects. Racialization describes the "extension of racial meaning to a previously racially unclassified relationship, social practice, or group."[9] Racialization describes how groups of people become defined as a race and how social and symbolic meanings become associated with said group. Racial projects bring together "what race means in a particular discursive or ideological practice" with how "social structures and everyday experiences are racially organized based upon that meaning."[10] The concept of racial projects account for the

---

[6] Michael Omi and Howard Winant, *Racial Formation in the United States,* 3rd ed. (New York: Routledge, 2015), 110.

[7] Omi and Winant, 111.

[8] Omi and Winant, 109.

[9] Omi and Winant, 111.

way that the concept of race is both about the cultural meanings associated with racial groups and about how social structures, such as the labor market, education, the law, and politics, are constructed in ways that unequally distribute resources along racial lines. Omi and Winant see racial projects as efforts made by both dominant groups and subordinated groups to simultaneously shape racial meanings and how social structures are racialized.

Omi and Winant see society as a collection of racial projects initiated by both dominant groups trying to maintain their social position and marginalized groups working to challenge social structures and cultural representations. They define racism as the collection of racial projects that work to "create or reproduce structures of domination."[11] Omi and Winant argue that because racialized groups are able to engage in racial solidarity and produce racial projects that resist racism, race "exceeds and transcends racism."[12] By defining racism narrowly as the collection of racial projects that specifically work to create or reproduce inequality, Omi and Winant are able to argue that race is more than racism.

Ronald Mize, while in agreement with most of Omi and Winant's theory of racial formation, disagrees with Omi and Winant's argument that race exceeds racism because he understands race as "the logical extension of how racism justifies, legitimates, and naturalizes unequal lived experiences."[13] Race is the product of racism created to naturalize racist ideas, actions, and social structures. Mize defines racism as "both a matter of interpersonal relations

---

[10] Omi and Winant, 125.

[11] Omi and Winant, 128.

[12] Omi and Winant, 128.

[13] Ronald Mize, "Critically Interrogating the Illusive Sign: The New 'Race' Theories and a Plausible Alternative for Understanding Cultural Racializations of Latino/a Identities." *Latino Studies* 11, no. 3 (September 2013): 342,

governing daily interactions and an institutional process that shapes the boundaries on large-scale social relations."[14] To Mize, Racism is both the daily interpersonal interactions that create and recreate racial lines and the social structures that govern a society. Mize argues that seeing race as a product of racism allows one to see racial identities as "imposed upon sub-groups" and see the construction of races as "embedded in specific unequal relations that require a rationale for explaining why particular privileged communities live, work, and even die in relative affluence."[15] I find Mize's argument compelling as it more clearly highlights that race is inextricably linked to racism, that we cannot discuss one without the other. I understand efforts to resist racism by building racial solidarity not as *exceeding* racism because if races are the product of racism, what meanings do racial categories have outside of racism?

For this book, I borrow from Mize (2013) to define racism as the structure of a society where social institutions, interpersonal interactions, and ideologies work in conjunction to unequally distribute economic, political, and social resources and rewards along racial lines. I draw from both Mize (2013) and Omi and Winant (2015) to define race as the product of racism used to justify and naturalize the unequal distribution of resources and rewards and as a concept that simultaneously refers to the cultural meanings associated with racial categories and the social position of racial groups within the social structures of a given society. These definitions guide my analysis of how race and racism are represented in DNA ancestry ads. Specifically, in these ads, race is represented not as a product of racism but as categories that describe one's biological genetic composition. By extension, racism in these ads is not the social structure of a society but prejudiced beliefs and actions based on a belief that people can be categorized into

---

[14] Mize, 359.

[15] Mize, 358.

mutually exclusive racial groups based on their ancestral background and that those racial groups are associated with certain behavioral characteristics. By obscuring the reality that race is a political category, these ads function to (re)create "biologically/genetically based categories of race" under the guise of scientific advancement.[16] These ads circulate ideas that reinforce the idea that race and racism are rooted in questions about human biology and not relations of power.

Media and Culture

Since I focus on the messages embedded within advertisements, I draw on theories regarding media and society from the British school of cultural studies. Scholars from the British school of cultural studies concerned themselves with the study of the culture of the British working class, which by the 1950s and 1960s led to an interest in how social movements and popular culture were being used to radically shift social ideas and attitudes towards class and culture.[17] Recognizing the power that popular culture, such as popular music, popular art, film, and television, had in shaping public opinion, scholars working within British cultural studies laid the foundation for the study of popular media within cultural studies.

A scholar central to the study of race and media is Stuart Hall. In "The Whites of their Eyes," Hall argues that the media's main purpose is the production and transformation of ideologies, which he defines as the images and concepts that provide the audience certain

---

[16] Michael Omi, "'Slippin' into darkness'": The (Re)biologization of Race," *Journal of Asian American Studies* 13, no. 3 (October 2010): 351,

[17] Jane Stokes, *How to Do Media and Cultural Studies* (Thousand Oaks: SAGE Publications, 2013), 41.

frameworks to interpret and make sense of the social world.[18] The relationship between racism and the media is that the media provides and circulates certain frameworks for interpreting and making sense of race and racism, specifically by constructing definitions of what race and racism are, the meanings each term carries, and the best approaches for combating racism. For this book, I interpret the advertisements I study as the avenue through which DNA ancestry testing companies circulate their frameworks of what race and racism are and their connection to genetics.

Hall also argues that any message that is communicated goes through a process of encoding and decoding.[19] In the process of encoding, the producer of a text embeds their message with cultural assumptions and frameworks of knowledge that shape how they understand the world. Once the text reaches an audience, the messages in the text are then decoded by the reader. Hall theorized that this exchange is not symmetrical and that readers can take three different decoding positions. Those who engage in a dominant reading interpret the text as the producer intended, which often means supporting dominant ideologies. Those who engage in a negotiated reading accept some of the propositions made by the producer of the text, while still questioning certain details. Those who engage in an oppositional reading recognize the producer's intent and choose instead to re-appropriate the message for counter-hegemonic purposes. This model allows me to see media as an institution that shapes peoples' perceptions

---

[18] Stuart Hall, "The Whites of Their Eyes," In *Gender, Race, and Class in Media: A Text-Reader*, 2nd ed, ed. by Gail Dines and Jean M. Humez (Thousand Oaks: SAGE Publications, 2003), 90.

[19] Stuart Hall, "Encoding/Decoding," in *Critical Theory: A Reader for Literary and Cultural Studies*, ed. by Robert Dale Parker (New York: Oxford University Press, 2012), 804-811.

about the world while acknowledging that audiences can interpret messages in media in different ways.

When examining the advertisements, I will explore the historical discourses about race, genetics, nation, and ancestry that shape how DNA ancestry testing companies produce their messages in their marketing and consider the implications of audiences accepting a dominant reading of these texts. Since I will not be exploring how audiences are interpreting the advertisements in this study, I cannot say much about the negotiated or oppositional readings audiences might take to these texts. Instead, I use Hall's model to acknowledge that the same text can produce various interpretations and that my reading of these texts is one of these many interpretations.

DNA and Society

My study will be in conversation with scholars who focus on the practice of genetic science and how geneticists' findings are interpreted and circulated throughout society. Scholars working on the relations between genetics and society understand science and culture as always being in conversation with one another, and they seek to explore how cultural assumptions and discourses influence scientific studies and how scientific findings influence society.

Dorothy Roberts in *Fatal Invention* explores how scientists, politicians, and business executives have repackaged older biological essentialist ideas of race under the veneer of 21st century scientific objectivity.[20] Pharmaceutical companies that claim they are able to produce race-specific medications, such as with the case of BiDil which was marketed as a heart medication developed for Black patients, present themselves as engaging in cutting-edge

---

[20] Dorothy Roberts, *Fatal Invention: How Science, Politics, and Big Business Re-create Race in the Twenty-first Century* (New York: The New Press, 2011), x.

research that is advancing the personalization of medical research when in actuality their findings are based on flawed studies with researchers going into the study already believing that there must be some level of biological similarity among people classified under the same racial group. DNA testing companies claim that the concept of genetic ancestry is completely divorced from the concept of race while simultaneously relying on biologically essentialist understandings of community and geography to build their databases, examine consumers' DNA, and present their results. Roberts worries that pharmaceutical and DNA ancestry companies being able to repackage the myth of a biological concept of race under the guise of cutting-edge science works to obscure the reality that race is, and has always been, a political category. Roberts' book serves as the basis for my understanding of the flaws of the science used by DNA ancestry testing companies and will aid me in thinking through how biological essentialist ideas of race are being presented in DNA ancestry testing advertisements under the guise of scientific objectivity.

Kim TallBear in *Native American DNA* outlines some of the social and political consequences the rise in DNA testing can have on indigenous communities. TallBear examines how geneticists, anthropologists, DNA testing companies, and genealogists construct the concept of "Native American DNA" and how they use this concept in their marketing and research.[21] TallBear argues that DNA testing companies rely on old biologically essentialist notions about indigenous peoples as well as a lack of knowledge about indigenous peoples and tribal sovereignty in order to construct the concept of "Native American DNA" which homogenizes hundreds of different communities. TallBear warns that the proliferation of the idea of "Native American DNA" and DNA ancestry testing could challenge tribal sovereignty as the right to

---

[21] Kimberly TallBear, *Native American DNA: Tribal Belonging and the False Promise of Genetic Science* (Minneapolis: University of Minnesota Press, 2013), 2-4.

define who is Native is taken away from Native communities themselves and given to DNA testing companies. TallBear's work further develops my knowledge on how DNA tests work and serves as a reminder of the real social and political consequences of these DNA tests.

A part of TallBear's argument expands on Donna Haraway's concept of "gene fetishism," a concept which I find helpful for this book as well. In *Modest_Witness@Second_Millenium*, Haraway defines gene fetishism as "mistaking heterogenous relationality for a fixed, seemingly objective thing."[22] Similar to the concept of commodity fetishism, Haraway states that gene fetishism involves forgetting "that bodies are nodes in webs of integration."[23] Emphasizing the gene as the most valuable source of knowledge about the self involves erasing that bodies and identities are constituted within various social relationships, historical processes, and relations of power. TallBear uses gene fetishism to describe the way that "Native American DNA" defines being indigenous as having to do with having "Native DNA" as opposed to being a social and political identity tied to history and one's position in the social structure of the United States.[24] I will use Haraway's concept of gene fetishism to explain the way that DNA testing companies' advertisements remove racial, ethnic, and national identities from their social and political context.

Past Studies on DNA Testing Marketing

Scholars in communication studies have looked at how DNA ancestry testing companies market their product to consumers and the assumptions about the relationship between genetics

---

[22] Donna J. Haraway, *Modest_Witness@Second_Millenium. FemaleMan_Meets_Oncomouse: Feminism and Technoscience*, 2nd ed (New York: Routledge, 2018), 142.

[23] Haraway, 142.

[24] TallBear, *Native American DNA*, 70-71.

and ancestry that these companies rely on and leave unquestioned in their marketing. Scodari (2017) conducted a qualitative analysis of AncestryDNA and 23andMe commercials from 2017 to examine whether biological kinship takes precedence over other forms of kinship, how hybridity is represented, if race and racism are facilitated or hindered, and whether consumers can form a resistant reading of these ads.[25] Scodari argues that the significance of biological kinship over other forms of kinship is the basic premise undergirding genetic testing marketing. Hybridity in these ads is commodified which results in a fetishization of difference based on acquiring objects, such as specific articles of clothing or pieces of furniture, that signify ethnic origins over a critical examination of the historical events that created the hybridity. Scodari argues Black consumers could develop a resistant reading by using these tests to link themselves to their African ancestral origins, however by doing so, Black consumers could inadvertently allow DNA ancestry testing companies to capitalize on Black Americans desire for a connection to a pre-slavery past. Finally, Scodari concludes that by relying on unquestioned assumptions of race and ethnicity as biologically essential and identifiable through DNA, DNA testing companies are complicit in the process of racialization and in the misuse of genetic science for racist aims.

Putman and Cole (2020) analyzed the rhetoric of AncestryDNA ads, which are Ancestry ads specifically about their DNA tests, from 2015 to 2018 to understand how Ancestry uses their ads to construct a symbolic need that consumers are then compelled to fulfill through the purchase of a DNA test. Using the theory of constitutive rhetoric as the framework for their analysis, Putman and Cole argue that AncestryDNA ads create a collective identity for the

---

[25] Christine Scodari, "When Markers Meet Marketing: Ethnicity, Race, Hybridity, and Kinship in Genetic Genealogy Television Advertising," *Genealogy* 1, no. 22 (2017): 2,

audience to identify with and suggest that the audience needs to take one of their DNA tests to reap the benefits of this new collective identity. The ads tell the audience that they are all carriers of DNA, which is constructed as being an objective source of knowledge about themselves that is waiting to be unlocked. Other ways of knowing about oneself and one's past, such as family narratives, are seen as subjective and less reliable sources of knowledge and the audience is compelled to go on a journey of discovery to find ones "rightful genetic community."[26] The audience, through the use of testimonials, are told to receive their results as if they are unquestionable and to act on them by embracing their newly discovered ethnic ancestry. However, this engagement is limited to interactions with certain objects, such as food, clothing, and language, and a detachment from the social and political consequences of embracing a new ethnicity. The framing of DNA as an objective source of knowledge serves to overlook the scientific critiques of DNA ancestry testing and promotes the notion of ethnicity as unproblematically biological. In this book, I add to these prior works by analyzing the assumptions present in themes I have identified in DNA ancestry tests not addressed in these previous studies: the theme of travel based on one's DNA test and the theme of DNA tests serving as tools that can fight against racism.

Methods

I conducted a content analysis of a collection of advertisements from Ancestry, 23andMe, and MyHeritage DNA where I identified common themes present in the ads. Afterwards, I grouped the ads based on these themes and conducted a more detailed textual analysis of ads in two specific themes that were not discussed in previous studies: the theme of ancestry travel and

---

[26] Putman and Cole, "All hail DNA," 231.

the representation of populations and regions outside of the U.S. and the representation of race, ethnicity, and racism and how DNA tests are touted as tools to fight against racism.

Content Analysis is a method for analyzing texts that involves "summing the occurrences of specific phenomenon such as specific words or images."[27] I began by searching for advertisements from Ancestry, 23andMe, and MyHeritage DNA on both YouTube and i.Spot.tv, a website that houses a database of television advertisements. I limited my search for ads about each companies' genetic ancestry DNA test since some of the ads were geared toward some of their other products, such as the health DNA test in the case of 23andMe and Ancestry or the family tree databases in the case of Ancestry and MyHeritage DNA.

This search resulted in a list of 53 unique advertisements: 40 from AncestryDNA, 10 from 23andMe and 3 from MyHeritage DNA. In an excel sheet, I jotted down each advertisement, and in a separate column I kept track of the words, images, and sounds present in the ad. I paid attention to how the concepts of race, ethnicity, nation, ancestry, family, and geography were being referred to both explicitly and implicitly. In a separate column for each ad, I created a list of themes based on the messages and imagery present in the ad. Through this search I was able to identify a few common themes. I then grouped the advertisements based on these themes.

After grouping the advertisements based on common themes, I conducted a textual analysis of the ads related to two themes not thoroughly discussed in previous studies: 1) the theme of ancestry travel and the representation of populations and regions outside of the U.S. and 2) the representation of race, ethnicity, and racism and how DNA tests are touted as tools to fight against racism. The theme of ancestry travel and the representation of populations and

---

[27] Stokes, *How to do Media and Cultural Studies*, 132.

regions outside of the U.S. emerged from advertisements where the focus of the ad is on the travel that a person took because of their DNA ancestry test or where populations and regions outside the U.S. are shown in relation to a person's DNA ancestry test results. The theme of the representation of race, ethnicity, and racism and how DNA tests are touted as tools to fight against racism emerged from ads that focused on outlining the relationships between genetics, race, ethnicity, and racism and how DNA test can be used to challenge racist ideas.

Alan Mckee (2003) describes textual analysis as "making an educated guess at some of the most likely interpretations that might be made of a text."[28] For McKee what makes a guess educated is that the interpretation is based on the researcher's knowledge about the culture where the text is circulated and is aware of the various intertexts that an audience would be familiar with, including "other texts in the series, the genre of the text, intertexts about the text itself, and the wider public context in which the text circulated."[29] I engaged in a textual analysis of the ads in the previously mentioned themes in an effort to provide an interpretation of how the messages in these advertisements might be received by the public. The intertexts I relied on where my understanding of historical and contemporary associations between race, ethnicity, and genetics in American culture. For each of these ads, I analyzed the dialogue, text, images, and sound to understand the assumptions supporting the claims made in the ad and the potential social and political implications of the claims made.

---

[28] Alan Mckee, *Textual Analysis: A Beginner's Guide* (London: Sage Publications, 2003), 70.

[29] Mckee, 93.

Positioning Myself

My investment in this study stems from my unease with how often I have seen people uncritically accept results from their DNA ancestry test and use those results to then change how they identify racially and ethnically. The uncritical acceptance of DNA ancestry test results worries me because it implicitly reinforces the idea that there is a biological component to the concepts of ethnicity and race.

As a scholar of color committed to crucially examining the structures that maintain racism in American society, the potential reinforcement of biologically essentialist ideas in DNA testing advertisement could prevent people from seeing that race has always been a political category that we have come to believe has a biological reality. While many believe that we have moved away from a seeing race as a biological reality and we now see it as a social construct, the popularity of genetic ancestry testing shows us that we have never completely given up the idea that distinct and observable biological difference exists among human beings. While seemingly benign at first, the reinforcement of a biological notion of race and ethnicity can result in people downplaying or overlooking the reality that identifying as certain races or ethnicity carries social and political implications and that racism can only be eliminated through political and social change. As a result, I come to the project with a focus on the wider implications the messages present in DNA ancestry test marketing can have on communities of color and their attempts to fight against the structures that maintain racism.

Chapter Breakdown

Chapter 1 offers an overview of the origins of biologically essentialists notions of race. This is followed by a review of the discourses about race, ethnicity, nation, genetics and ancestry that inform how DNA ancestry testing companies collect their data, interpret their data, and represent their results. It is the knowledge about these discourses, as well as the limitations of what the tests can prove, that inform my analysis of the advertisements in the following chapters.

In chapter 2, I outline how ancestry travel is represented in a collection of advertisements. I unpack how consumers are meant to understand the relevance of traveling to the places from where their ancestors supposedly came. I address how traveling to re-connect with one's roots becomes commodified and how interactions between travelers and the people they visit are represented as transactional and focused on what people around the world can offer the travel on their journey of self-discovery. I also analyze ads where images of different regions around the world are shown as a person discusses the results of their DNA ancestry test. While the person given the testimonial may not have engaged in travel themselves, these ads provide an example of how DNA ancestry testing companies represent the different "reference populations" they use in their tests in their advertisements.

Chapter 3 focuses on how race, ethnicity, and racism are defined in these ads and how DNA ancestry tests are touted as tools in the fight against racism. I analyze the way these ads represent the relationships between genetics, race, ethnicity, and racism, showing how these concepts are often conflated with one another and how race and racism are rooted in the biological. DNA tests are shown as potential solutions to racism as it proves that race does not actually represent the biological reality of human diversity. This understanding of the root causes of racism erases the fact the race is a political concept related to both social structures and representation and that ending racism can only come from political struggles.

The conclusion includes my final reflections on the social and political consequences DNA ancestry testing have already had and the potential consequences it could have in the future, I will also provide my thoughts on future directions of research.

CHAPTER I: THE SCIENCE OF DNA ANCESTRY TESTS

Introduction

In 2017, DNA tests were everywhere. From TV ads and online ads to displays at local pharmacies and retail stores, it seemed as if I was seeing DNA tests wherever I went. In fact, it was seeing a large display for 23andMe tests at my local wholesale store back in 2017 that sparked my initial interest in people's fascination with DNA ancestry tests. The main claim that DNA ancestry testing companies make is that a DNA test will be able to provide a test taker with information about their ethnic ancestry through an analysis of their DNA. However, how exactly is a DNA test able to provide such information? How do these tests define certain pieces of DNA as Spanish or West African? Also, do not these tests veer dangerously close to suggesting that there are biological components to ethnicity, nationality, and even race? I use this chapter to answer these questions through a literature review of scholarship related to DNA and DNA testing.

My analysis of DNA ancestry testing advertisements and marketing is influenced by the work of biologists, anthropologists, ethnic studies, and science and technology studies scholars who have critiqued the assumptions underpinning the science behind DNA ancestry tests and warned of the potential consequences the acceptance of the results offered by these tests could have on how the public understands the relationship between biology and race. I use this chapter to explain what DNA ancestry tests are, what they test for, the assumptions embedded in the process of their creation, and what a consumer's test results actually mean. Additionally, I will explore how these tests are built on a history of racial science, spanning from the enlightenment to the present, as well as how these tests are engaging in a redefinition of race.

I borrow the term "racial science" from sociologist Dorothy Roberts, who finds it to be is a more accurate descriptor for the customary use of science to perpetuate a biological understanding of race throughout the last three centuries in contrast to the often used phrase "scientific racism" which implies an "exceptional use of science to create and support racist ideas."[30] The assumption that racial categories reflect some degree of biological similarity among those in the same racial group and dissimilarity between those of different groups has shaped how scientists have gone about learning more about the world, from the naturalists to today's genetic scientists. This history informs the science behind the construction of DNA ancestry tests. Therefore, I begin the chapter outlining the history of racial science from the enlightenment to the present, highlighting along the way how society has shaped science and how science has shaped society.

After that, I will turn to DNA ancestry tests specifically, explain the types of tests that are being sold and the assumptions that inform their creation and results offered to the consumers. Here I will show how questionable science and commercial interests combine to give consumers a false sense of what these tests can tell them about their past. I will end by exploring how DNA ancestry tests are redefining race in the present moment, the consequences this redefinition can have for communities of color, and how the information in this chapter informs my analysis.

The History of Racial Science

The sciences have always been in conversation with popular societal beliefs and assumptions at any given time. Societal beliefs have shaped what questions scientists have asked and the methods they have employed, and scientific findings have been used to validate societal beliefs about the nature of the world. Racial science emerged at the confluence of two major

---

[30] Roberts, *Fatal Invention*, 27.

social processes, the enlightenment and the age of European colonialism.[31] The enlightenment

marked a move away from relying on religious explanations about the world to engaging in

secular systematic analyses of nature to uncover the laws governing society. Natural science

developed alongside European colonialism. As European nations began building their empires

across the world, they interacted with peoples of other continents and needed to build systems to

manage the new populations they governed and to maintain their power over them. Scientific

ideas of race developed at this time because scientists sought a non-religious explanation for the

human diversity they encountered and because if the concept of race represented a natural and

biological division of humanity, race could be used to suggest that the prevailing colonial order,

which involved a race-based slavery system, the expulsion of indigenous peoples, and

subjugation of racialized peoples in their colonies, was a natural consequence of innate

biological differences.[32]

Naturalists were the first to make race a scientific object of study by applying the method

of taxonomy, which they used to study plant and animal species, to the study of humans.[33] The

science of taxonomy involves naming and organizing living things into different groups based on

biological traits they share with one another. Naturalists believed that the diversity in human

appearance that they observed represented natural divisions between groups of human beings

like the divisions they observed in other animal and plant species. Naturalists decided to group

human beings based on these observable characteristics, and this resulted in the first major

---

[31] Johnathan Marks, "Race: Past, Present, and Future," in *Revisiting Race in a Genomic Age*, ed. Barbara A. Koenig, Sandra Soo-Jin Lee, and Sarah S. Richardson (New Brunswick: Rutgers University Press, 2008), 21.

[32] Omi and Winant, *Racial Formation in the United States,* 115.

[33] Roberts, *Fatal Invention*, 28.

scientific idea about race that still shapes how race is thought about today: that "races" represented "a fundamental taxonomic division of the human species."[34] This meant the people in the same racial group were more biologically similar to one another then they were to people in different racial groups.

The idea that human beings belonged to different races became the predominant explanation for human difference and scientists in the 18th and 19th century engaged in various studies to figure out the biological features that distinguish one race from another. Scientists engaged in extensive studies of the body, from skin color to hair texture, nose size, skull size, and brain capacity all in an attempt to "delineate the biophysical markers of race" and categorize individuals and their bodies within these racial categories.[35] Scientists also began to argue that observable differences between humans were tied to innate differences in complex human characteristics, such as intelligence, and this association intensified with the advent of eugenics in the 20th century.

One of the major debates among scientists that emerged at this time was that between the theories of monogenesis and polygenesis.[36] Those who believed in monogenesis argued that all humans shared the same origin while those who believed in polygenesis argued that human beings had different origins and did not belong to the same species. Polygenesis offered Europeans and white Americans a convenient justification for the dominant political and social order because it suggested that groups of human beings were fundamentally different from one another and that some humans were biologically superior to others.

---

[34] Marks, "Race: Past, Present, and Future," 22.

[35] TallBear, *Native American DNA*, 35.

[36] Roberts, *Fatal Invention*, 31.

Charles Darwin's *On the Origins of Species* is seen as the text that put the Monogenesis vs. Polygenesis debate to rest. Darwin lays out his theory of evolution that all living organisms have evolved over a long period of time through the process of natural selection and argues that all humans share the same common ancestor. While Darwin's theory effectively ended the monogenesis-polygenesis debate, the idea of a racial hierarchy that stemmed from the theory of polygenesis was reinterpreted through an evolutionary framework.[37] Different races were seen as representing different stages of human evolution, with Europeans being the most evolved humans, and the theory of natural selection was reinterpreted to suggest that in the struggle between races, Europeans came out on top because they were the "fittest."[38]

Racial science eventually led to the emergence of eugenics, which was popularized by Francis Galton.[39] Eugenicists reinterpreted Medellin genetics to claim that many complex human behavioral traits, such as intelligence, artistic ability, and work ethic, were genetically determined and inherited from one's parents. Eugenicists used their theories to argue that the various disparities and struggles faced by racialized groups were a result of their biology as they did not inherent traits that would help them succeed and in fact had inherited traits that impeded their success. Any social programs made with the intention of eliminating these disparities, they argued, would be a waste of resources. Eugenicists believed that the only way to solve social problems was though selective breeding.[40] The government needed to stem the reproduction of those with undesirable traits and promote the reproduction of those with desirable traits.

---

[37] TallBear, *Native American DNA*, 36.

[38] TallBear, 36.

[39] Omi and Winant, *Racial Formation in The United States*, 116.

[40] Roberts, *Fatal Invention*, 36.

Eugenicist thinking influenced many social programs and policies that had disastrous consequences for racialized populations and poor white communities, such as restrictive immigration policies and the forced sterilization of poor white women and women of color. This moment again highlights how the imperatives of dominant groups who hold power influenced the scientific questions that were investigated and the scientific findings that were given authority. Eugenicist thinking remained popular up to the middle of the 20th century. At that point, the horrors of the eugenicist program in Nazi Germany, the advent of the civil rights movement, and the emergence of the new field of population genetics, caused many to question the idea that "races" reflected a fundamental division of humanity and that complex human behavioral traits were biologically determined and inheritable.[41]

The field of population genetics suggested that the more appropriate way to study human genetic diversity was to study how people varied based on their geographical location. Instead of studying differences in races, they would study differences in populations. While many populations geneticists challenged the ideas that complex human traits were biologically determined that certain behavioral and mental traits were essentially tied to certain racial groups, and that there was an inherent hierarchy between human beings based on the category of race, the shift to theories of population genetics did not necessarily entail a discrediting of the idea of race altogether. In fact, the shift to population represented a change in ideas about race, but not a negation of the concept.[42] While against the idea that behavior or ability was tied to race, many still believed that humans could be divided taxonomically. In essence, what they took issue with

---

[41] Roberts, 46.

[42] Marks, "Race: Past, Present, Future," 23.

was that race was used to shape social policy and justify disparities. However, many still found it a useful biological concept to be used to conduct research.

The distinction being made here is evident in the UNESCO "Statement on Race and Race Difference" released in 1950. While the authors of the statement made it clear that all humans share a common origin and that there are more similarities than differences between human beings, they still saw race as a meaningful biological concept. They make a distinction between race as a biological concept, defined as "one of the group of populations constituting the species Homo Sapiens" that exhibit certain physical differences as a result of their historic isolation from on one other, and race as a social concept where a race is simply "any group of people whom they choose to describe as a race."[43] They state that anthropologists at the time of writing divide humans into three major human races: Mongoloid, Negroid, and Caucasoid. They also argued that equality as an ethical principle should not depend on human beings being "equal in endowment" and that biological differences that exist between ethnic groups should not play a role in the social and political organization on society.[44] It is clear from this statement that the main concern at the time was not that people believed in the concept of race, but that they misunderstood and misused it. Scientists still believed that there were significant biological differences between different populations that could be described using the term "race." They just believed that the public was misusing the term and overexaggerating the difference between groups of people to shape social policy, which had disastrous consequences.

---

[43] UNESCO, "Statement of on the Nature of Race and Race Difference 1950," In *Ethnic Studies: Critical Fundamentals*, ed. Tim Messer-Kruse (Toledo, OH: Achromous Books, 2018), 95.

[44] UNESCO, 96.

From this point forward, a split began to form between different groups of population geneticists and biological anthropologists. One group of scientists used their science to argue the concept of race had no biological basis and should no longer be use as a meaningful biological category. Another group of scientists still wanted to hold on to the ideas signified by the concept of race, specifically that people within the same racial group had to share some genetic similarities since their ancestors were imagined to share a particular origin (usually on a continental level) and that the relatedness within people of the same racial group had to be higher than that between people of different racial groups.[45] Scientists in this latter camp decided to move away from using "race" since it had become a controversial term, but held on to the assumptions embedded in the concept. These scientists tried to create an imaginary wall between science and society, claiming that their work was trying to uncover the objective genetic differences between people that exist under the skin and that it was completely separate from racial politics.[46] By holding on to the belief that we can identify and divide humans into discrete genetic clusters, many scientists have reconfirmed old ideas of race within the new categories of "ancestry" and "population" under the guise of 21st century scientific objectivity.

The New Racial Science

The first term that emerged to replace the concept of race in the late 19th century was the term "population" which was used to define a group of people said to share many genetic similarities with one another. The move to population was seen as more accurate description of human genetic variation since scientists studied how human groups varied genetically because of geographic distance from one other instead of making assertions about genetic relatedness based

---

[45] Roberts, *Fatal Invention*, 47.

[46] Roberts, 47.

on racial categories. The term "ancestry" came later on and was used as a way to acknowledge that members of a group that share genetic similarities are likely to pass down that a set of traits to their descendants while using "race" to signify a political category.[47] While used differently, these terms are fundamentally related since "ancestral groups" are imagined as populations who shared genetic similarities who at one point were isolated from other populations, but within the last few centuries have intermixed with other populations and as a result their genetic traits are joined with that of other populations. Therefore, while emerging at slightly different times, the two terms, population and ancestry, are related and are often used interchangeably.

Important insights about human genetic variation have come from the fields of population genetics and biological anthropology that provide evidence that race, as a theory of how human biological difference is patterned, is inaccurate. First, from these fields of study we learn that human variation follows a clinal distribution, which means the frequency of certain genetic traits vary gradually across geographical space moving from one extreme to another with few breaks in between.[48] Additionally, any population is genetically similar to populations that are found nearby, and this genetic similarity is inversely related with distance. The farther apart populations are the less genetically similar they are. Since genetic similarity is relative to the location of the population in question, the belief that populations within a continent are more similar to each other than those in another continent, which serves as the basis for the argument that racial categories are accurate proxies for genetic similarity, ignores the clear divisions

---

[47] Roberts, 63.

[48] Deborah A. Bolnick, "Individual Ancestry Inference and the Reification of Race as a Biological Phenomenon," in *Revisiting Race in a Genomic Age*, ed. Barbara A. Koenig, Sandra Soo-Jin Lee, and Sarah S. Richardson (New Brunswick: Rutgers University Press, 2008), 72.

between continents cannot exist since populations living at the edge of these continents will be more genetically similar to each other than populations within their respective continents.

Another insight is that most genetic diversity in the world is found within the continent of Africa and that there is more genetic diversity between African populations then there is between non-African populations.[49] These findings reflect the evolutionary history of human beings. Since Homo Sapiens first evolved in Africa, African populations have had the longest time to develop genetic differences. As human populations began to migrate out of Africa, they carried with them only a subset of that genetic diversity and each subsequent migration resulted in a subgroup of humans more genetically similar to each other than those who remained in Africa.[50] This data provides important evidence for the inadequacy of the concept of race for describing human genetic variation. Some scientists have used these findings to argue that any use of race is dangerously reductive and should no longer be used. On the other hand, there are some who believe the concept of race serves as a helpful approximation of how genetic human variation is structured while acknowledging its limitations. It is this debate about the viability of using the concept of race that has kept the door open to race being redefined in the concepts of population and ancestry.

As Professor of Human Geography Catherine Nash outlines in *Genetic Geographies: The Trouble with Ancestry*, there are three types of studies in the field of population genetics that have left the door open to the rearticulation of race. The first are studies that treat the genetics of culturally defined human groups as their object of analysis.[51] These studies are interested in

---

[49] Bolnick, 71.

[50] Bolnick, 72.

learning more about the history of human genetic variation and the evolutionary history of, and relationships between, various human groups. For these studies, researchers often begin by defining a set of culturally and linguistically defined groups they would like to study and then they go about collecting samples of DNA from these groups. Studies focused on human evolution tend to see indigenous communities on the various continents as the most representative of human genetic variation in the premodern era. These communities are imagined as having been relatively isolated from other groups prior to the era of colonialism. These studies begin by picking indigenous communities across the globe as their object of analysis and then collect a sample of genetic material from these communities.

This process of study design and data collection highlight a few issues with how the concept of population is used in practice. First, considering that there are no clear dividing lines in genetic variation between humans, for populations to be meaningfully studied in this way they need to be imagined as isolated "gene pools" that are static, or at the very least, have remained mostly unchanged over a long period of time.[52] Currently living indigenous communities are imagined as having lived the longest time as geographically isolated communities who only recently started mixing with other populations in the modern era as a consequence of colonialism. This understanding of populations privileges a specific moment of contact that erases the reality that human groups have always been migrating and interacting with each

---

[51] Catherine Nash, *Genetic Geographies: The Trouble with Ancestry* (Minneapolis: University of Minnesota Press, 2015), 37.

[52] Lundy Braun and Evelynn Hammonds, "The Dilemma of Classification: The Past in the Present," in *Genetics and the Unsettled Past: The Collision of DNA, Race, and History*, ed. Keith Wailoo, Alondra Nelson, and Catherine Lee (New Brunswick: Rutgers University Press, 2012), 68.

other.[53] Thus, to imagine contemporary indigenous people as historically isolated gene pools immobilizes them historically and erases the genetic diversity that exists in these communities because of historical migrations and intermixing.

This highlights a second issue, which is that by starting with culturally defined human groups and using them to define the populations that will be studied genetically, an association is made between genetics and cultural difference.[54] Some scientists argue that regardless of how carefully those conducting the studies are, by basing the populations chosen for these studies on cultural differences, these studies seem to suggest that the boundaries between sociocultural and political communities reflect genetic boundaries, which often is not true. Related to this is the third issue, which is that that culturally defined populations are not a given; in fact populations are "actively produced social formations" that have been created and recreated over time as a result of historical processes and scientific studies.[55] Braun and Hammonds highlight how linguistic and social anthropologists, through academic knowledge production, have transformed fluid African societies into sharply demarcated discrete entities "suitable for scientific definition, sampling, and classification."[56] Roberts provides the example that ethnic groups like the Yoruba and Mende that are not groups constituted of people who are genetically similar, but groups created out of the political forces arising out of European colonialism.[57] Treating these ethnic groups as one genetically defined populations flattens the cultural and genetic diversity of these

---

[53] Braun and Hammonds, 78.

[54] Nash, *Genetic Geographies*, 37.

[55] Braun and Hammonds, "The Dilemma of Classification," 78.

[56] Braun and Hammonds, 68.

[57] Roberts, *Fatal Invention*, 248.

groups and implies a one to one relationship between cultural and linguistic difference and genetics.

The second group of population genetics studies that open the door to the rearticulation of race are those that focus on what they call "admixed" populations, or populations of mixed ancestry.[58] The goal of these studies is to learn more about the relative genetic contributions of certain ancestral populations in admixed populations of interest. Admixture studies have been conducted by biomedical geneticists in an effort to discover genetic explanations for the observed difference in prevalence of certain disease between racial groups. However, the connections made between genetics and race in these studies have important implications for other forms of genetic science. In fact, admixture mapping is the method of genetic analysis that is used to develop one of the major types of DNA ancestry tests.

Rajagopalan and Fuijmura offer a detailed explanation of how admixture studies are conducted, and the ways race get reproduced within them. Those who conduct admixture studies believe that the differences in prevalence of certain diseases across racial lines might be explained partially by genetic traits that those in particular racial groups might share. In these studies, researchers use the social category of race to infer high levels of relatedness among members of a particular race category and lower levels of relatedness between members of different racial categories.[59]Admixture studies study populations believed to be "genetically admixed" by which they mean those who descend from two ancestral populations, believed to

---

[58] Nash, *Genetic Geographies*, 38.

[59] Ramya Rajagopalan and Joan H. Fujimura, "Making History via DNA, Making DNA from History: Deconstruing the Race-Disease Connection in Admixture Mapping" in *Genetics and the Unsettled Past: The Collision of DNA, Race, and History*, ed. Keith Wailoo, Alondra Nelson, and Catherine Lee (New Brunswick: Rutgers University Press, 2012), 145.

have previously been geographically isolated, who started intermixing within the last 20 generations.[60] The idea behind the 20 generation limit is that the "mixing" is relatively recent enough that large identifiable parts of one's genome were inherited virtually unchanged, which make them easier to identify in a genetic analysis. The racial groups in the U.S. that are classified as "admixed" are Black and Latinx Americans who are understood as descended from at least two ancestral populations. In this context, the ancestral populations are often defined at the continental level and chunks of DNA are identified as representing "continental lines of descent" such as African or European.[61] Black Americans are understood as descended from Africans and Europeans while Latinxs are understood as descended from Europeans, Native Americans, and Africans.

A central aspect of admixture studies is the construction of Admixture maps. Admixture maps rely on identifying "Ancestry Informative Markers" (AIM) which are segments of DNA that are said to be representative of certain ancestral populations. Between human beings, there are small segments of a DNA that vary from person to person. These small segments are called "single Nucleotide polymorphisms" or SNPs for short. An example of a SNP would be that at a particular location in a genome, some people will have the DNA base "A" at that spot while others might have the DNA base "C." Geneticists observe differences in the frequencies of certain SNPs between populations to learn more about the genetic composition of different populations. AIMs are basically a subset of SNPs that are defined as informative for ancestry.

To identify AIMs, Geneticists take an existing genetic database that has DNA samples from contemporary populations across the world and divide the data into groups based on their

---

[60] Rajagopalan and Fujimura, 146.

[61] Rajagopalan and Fujimura, 147.

continental origin. Then, they compare two of these continental populations with one another and find the difference in frequency of specific SNPs within said populations. Most SNPs appear almost in equal frequency in various populations. Therefore, the SNPs defined as AIMs are the very few SNPs that have a high difference in frequency between the continental level populations.[62] To create an admixture map for an admixed population, researchers define what the two ancestral populations of the admixed population are and then compare the SNPs found in the two ancestral populations and define the ones with the largest variants between the ancestral populations as AIMs. For example, in the case of Black Americans, the ancestral populations are defined as "European" and "West African" because Black Americans in the present are assumed to be descendants of the Europeans and African who came into contact during the age of European colonialism and the creation of the institution of slavery. Admixture mapmakers compare the difference in frequencies of particular SNPs in the "European" and "African" populations and those with the largest differences are defined as AIMs. The AIMs that appear in higher frequency in the "West African" group are defined as AIMs for "West African" ancestry while those that appear with higher frequency in the "European" group are considered AIMs for "European" ancestry. This list of AIMs is then used to analyze the DNA of Black Americans and SNPs found in this population are defined as either indicative of either African ancestry or European ancestry.

One of the major issues with admixture mapping is that populations are often defined at a continental level. Not only does that flatten the genetic diversity that exists within populations in a continent, but it also maps onto the older continental definition of race.[63] Race was, and still

---

[62] Rajagopalan and Fujimura, 150.

[63] Nash, *Genetic Geographies*, 39.

often is, associated with geographic origin and that stems from the naturalists who divided people into types based on origin. Since grouping people based on their continental origin is not an accurate estimate of how genetic diversity exists in the world, that scientists continue to assume that people from the same continent are bound to be more genetically similar to one another than they would be to people on another continent shows how cultural assumptions inform how science is done. The act of naming continental groupings "populations" in these admixture studies also shows that the concepts of populations and race are not as distinct as they are often thought to be. In fact, the effective interchangeability of the two implied by this practice can lead to a reconfiguration of "race as genetic ancestry."[64]

Another issue is that while scientists use the term "ancestral population," the DNA that they are using is not from people who lived hundreds of years ago in relative geographical isolation from one another. Instead, they analyze DNA from contemporary populations living in those continents and infer that their DNA is directly descended from the people who lived on those continents hundreds of years ago. Using this assumption, geneticists use the DNA of contemporary populations as a stand in for DNA of the so-called "ancestral populations" that contributed to the contemporary admixed U.S. populations.[65] Rajagopalan and Fujimura write that in an admixture study involving Black Americans, for the "West African" ancestral population they used DNA from contemporary West Africans and for the "European" ancestral population they used DNA of contemporary white Americans. By treating currently living West Africans as a stand in for ancestral West Africans whose DNA was inherited by contemporary Black Americans, these studies treat living Africans as if they are relics of the past, as if their

---

[64] Roberts, *Fatal Invention*, 63.

[65] Rajagopalan and Fujimura, "Making History via DNA," 151.

DNA has remained unchained from prior to the beginning of European colonialism to the present. [66] It also erases the mixing and migrating that happened during that period let alone that there is so much genetic diversity in Africa that an "African" ancestral population is basically nonsensical. In the case of the "European" DNA, they treated contemporary white Americans as a stand in for European ancestors. Similar to the issue with the use of West Africans, the genetic diversity among populations in Europe and intermixing with peoples in other continents before colonial contact is erased. Additionally, if the only selection criteria is being a "white" American, then whiteness is treated as a transhistorical category where people considered white in the present are assumed to be genetically similar to the European populations that arrived in the Americas in the 16th and 17th century. The use of "white" Americans as a stand in for ancestral Europeans also takes for granted the idea of a fixed and pure whiteness that has existed since Europeans first set foot in the Americas.[67] Since admixture mapping plays a large role in Ancestry DNA testing, the issues presented by these studies are also present in DNA ancestry tests.

Finally, the third group of population genetics studies are those interested in better understanding the global structure of human genetic variation. [68] Scientists conducting these studies are interested in using statistical tools to see if there are any better ways to categorize human genetic variation. These studies often take it as a given that race categories serve as accurate proxies for how human variation is structured. The assumption that race represents accurate groupings of people by genetic similarity shapes how these studies are conducted and

---

[66] Braun and Hammonds, "The Dilemma of Classification," 78.

[67] Rajagopalan and Fujimura, "Making History via DNA," 152.

[68] Nash, *Genetic Geographies*, 40.

how the findings from these studies are represented to the public by scientists and scientific reporters.

A comprehensive study of global human variation would involve a systematic random sampling of people from across the global, however this is very time consuming, expensive, and raises issues of ethics since not all groups would agree to having their DNA sampled. As a result, the databases collected for these studies usually involve collecting samples of smaller populations who are meant to stand in for larger populations. This practice raises questions about the scale at which a population is being defined and the selection criteria used to pick the samples meant to stand in for a larger population. Scientists who believe a continental level of analysis is an appropriate proxy for genetic diversity will pick samples of populations within a continent and have those people stand in for the genetic diversity of the entire continent. This practice works to present a flattened version of the genetic diversity present in a given continent.

An additional layer to this problem is that the populations selected to stand in for the continent are selectively chosen to emphasize difference.[69] Since human variation exists on a gradient, if populations that are geographically close but on two separate continents, such as say people in Spain and people in Morocco, they would appear very genetically similar. Knowing this, populations that exhibit a lot of intermixing between continents are purposefully excluded and those who are geographically the farthest apart are chosen to stand in the for the continent.[70] This again reinforces the idea that the populations in the continent are so genetically similar that they could just stand in for one another and forces the data, through data collection, to fit a continent level analysis. Therefore, assumptions about how global human variation "should"

---

[69] Marks, "Race: Past, Present, Future," 25.

[70] Roberts, *Fatal Invention*, 64-65.

look influence how the data is collected in the first place, so it is not a surprise that the results of these studies seemingly support these assumptions.

Another issue is these assumptions influence how the scientific findings are framed and how they are represented in the press. Bolnick writes about a couple of studies in genetics that have been referenced as providing genetic evidence for the existence of races. Both of these studies used a program called *Structure*, which is basically a statistical program that takes a dataset of DNA samples and tries to group them into genetically similar clusters based on the number of clusters the scientists tell the program to try to test. Researchers then test various numbers of clusters, and the program will offer scores of probabilities based on the likelihood that the given number of clusters is closest to representing naturally existing clusters of this DNA. Where concerns arise is that there have been some *Structure* studies that are credited as showing "proof" that races are accurate proxies for genetic similarity. For example, Rosenburg et al. (2002) claimed in their abstract that after testing multiple sets of clusters, they found that their data had one of the highest probabilities at being clustered into six main groups, five of which matched major geographic regions: Africa, Eurasia, East Asia, Oceania, America, and the Kalash of Pakistan.[71] Science writer for the *New York Times* Nicholas Wade, and many others in the press, wrote that this study proved that there was a biological reality to race since five of the six major groups outlined in the study aligned with popular ideas about race as tied to geographical origin.[72] After digging deeper into the study, Bolnick learned that the team had tested many more levels of clusters originally and even found just as high, if not higher, probabilities at larger clustering numbers, such as 10 or 20. However, while Rosenburg et al acknowledged this, they

---

[71] Bolnick, "Individual Ancestry Inference," 76.

[72] Roberts, *Fatal Invention*, 61.

decided to exclude those findings from the final report because they wanted to focus on results that would be much easier to replicate. They never made the claim that these findings "prove" that race is a valid proxy for genetic similarity and difference. Instead Wade and other science writers interpreted the findings in this way and decided to present it that way to the public, most likely because they easily fit within the popular belief that "continental groupings are biologically significant."[73] Therefore, just as the scientists' assumptions affect how studies are conducted and how data is collected, science writer's own assumptions influence how they reported scientific findings to the public, which can have dangerous implications for how the public understands the relationship between race and genetics.

It is clear then that biological ideas about race were never completely discarded, but instead were rearticulated through new technologies of science due to unacknowledged assumptions that influenced data collection and interpretation. These issues get heightened when they intersect with commercial influences to try to sell the benefits of genetics to an audience as is seen in the case of DNA ancestry tests.

DNA Ancestry Tests

There are two major types of DNA ancestry tests out in the market: parental and maternal linage tests and Ancestry Informative Markers (AIM) tests. The first type of tests, called lineage tests, examines DNA found in a person's mitochondria or Y chromosome in order to trace their maternal or paternal lineage.[74] DNA in the mitochondria and the Y Chromosome are considered

---

[73] Bolnick, "Individual Ancestry Inference," 77.

[74] Mark D. Shriver and Rick A. Kittles, "Genetic Ancestry and the Search for Personalized Genetic Histories," in *Revisiting Race in a Genomic Age*, ed. Barbara A. Koenig, Sandra Soo-Jin Lee, and Sarah S. Richardson (New Brunswick: Rutgers University Press, 2008), 203.

important because they are non-recombinant, which means that they are passed down from parent to child partially unchanged.[75] Most of the DNA in a person's genome is recombinant because for each of the pairs of chromosomes one inherits from their parents, one chromosome is inherited from each parent. Over many generations, the chromosomes in a person's DNA can be understood as a combination of various generations of ancestors, with each individual ancestor's contribution becoming smaller and smaller as generations pass. However, Mitochondrial DNA and the Y-Chromosome do not go through this process and mothers pass on their Mitochondrial DNA unchanged to their children and fathers pass on their Y-Chromosome unchanged to their sons. DNA tests build on this knowledge by attempting to use MtDNA and Y-Chromosome DNA to trace the maternal and paternal linage of the test taker back for upwards of 8 to 10 generations. Geneticists accomplish this by comparing genetic markers found in an individual's MtDNA or Y-chromosome DNA to their reference database of genetic markers that have been identified in specific populations and try to find a match.[76] These tests often give people answers either at the level of a continent or region in a continent, such as "Native American," "East Asian," or "West African." In cases where an individual has a rare genetic marker, these tests can be used to suggest that the individual might be able to trace their lineage back to particular ethnic groups or local level populations.[77]

Related to these types of tests are a process some DNA ancestry testing companies do called DNA matching.[78] In DNA matching, genetic testing companies compare the DNA of a

---

[75] Greenly, "Genetic Genealogy," 217.

[76] Shriver and Kittles, "Genetic Ancestry," 203.

[77] Shriver and Kittles, 204.

test taker to the DNA of all the test takers in their company DNA. Through this comparison they find people who share matching segments of DNA that they both inherited from a common ancestor. Once they find a match, they determine the degree of shared DNA to estimate how closely related the two test takers are, the closer the match the more closely related they are to one another. It is through this process that companies like Ancestry and 23andMe are able to identify potential cousins in their database. If they find a match, they inform both test takers who are each shown a file of the person the company believes they are related too and the estimated degree of relatedness.

The second type of DNA ancestry tests are called either Ancestry Informative Markers (AIM), Biogeographical Ancestry, or autosomal markers tests. In contrast to the lineage tests, these tests examine the parts of one's genome that is recombinant to make guesses about what ancestral populations contributed to one's genome.[79] These tests basically take the methodologies used in admixture mapping and apply them to guess a person's ancestry. The makers of these tests identify the various SNPs that exist in their databases and compare the frequency of various SNPs between different predefined populations, usually at the continental or regional level. They then strategically pick the SNPs with the highest degree of difference in frequency between various populations and label them "Ancestry Identification Markers" or AIMs. The AIMs that appear with higher frequency in a given population are labeled as AIMs for that population. A prime example is the Duffy Null allele which is a genetic trait that gives people an advantage against malaria that appears at a very high frequency in virtually all Sub-

---

[78] Ancestry, "What is DNA Matching?" Ancestry.com, Accessed July 13, 2021, https://www.ancestry.com/cs/dna-help/matches/dna-matching.

[79] Greenly, "Genetic Genealogy," 222.

Saharan African populations and rarely appears in populations outside of the region.[80] AIM ancestry tests take an individual's DNA sample and then compare the SNPs in that person's sample to the list of reference database of AIMs the company has and then makes a calculation of best guesses as to where the individual's SNPs came from and the percentage contribution of each ancestral population in the person's DNA. Using the Duffy Null Allele as an example, anyone with that AIM in their DNA sample would be classify as having at least some percentage of "Sub-Saharan African" DNA depending on how many other AIMs for that population are found in their DNA. The results the test taker gets is usually called their "biographical ancestry estimate" defined as the estimate of "a person's ancestry in terms of the proportional representation of AIMs from a selection of ancestral populations."[81] These estimates are presented to consumers as percentages that add up to 100%. Therefore, an individual test taker could get results back that state they are 50% Western European, 20 % Sub-Sharan African, and 30% Native American.

Limitations of DNA Ancestry Tests

Just as there are many issues with the use of race in new genetic science, these tests, based on those studies, also have many technical limitations, and rely on questionable assumptions.

The first assumption pertains to the lineage-based DNA tests. People most likely take these tests to learn more about their family history, with the appeal being that they will get an answer with high accuracy considering that the MtDNA and/or the Y-Chromosome are inherited practically unchanged from generation to generation. However, the drawback is that you get

---

[80] TallBear, *Native American DNA*, 72.

[81] Shriver and Kittles, "Genetic Ancestry," 205.

information about 1 or 2 lines of decent, or one ancestor per generation, out of 10+ generations of ancestors.[82] Considering that every generation, the number of one's biological ancestors doubles, (i.e., one has 2 parents, 4 grandparents, 8 great-grandparents) the test taker is getting an incredibly limited about of information about their ancestry. Moreover, there is the fact that it is quite possible that significant parts on one's lineage do not show up on a test. TallBear (2013) provides a hypothetical example that a test taker could have two Indigenous grandparents, one from their mother's side and their father's side, but if their indigenous grandfather was on their mother's side and their indigenous grandmother was on their father's side, neither would show up in the lineage tests.[83] The test taker's mother would have inherited their MtDNA from their non-indigenous mother and the test taker's father would have inherited their Y-Chromosome from their non-indigenous father. This example illustrates the limited amount of information these tests could offer which is often not made clear to the test taker.

When it comes to the AIM or autosomal markers test, while the appeal is that they test more parts of the genome and can therefore hypothetically capture more information about one's ancestral lineage, these tests run into similar problems faced by admixture mapping. The first issue is related to how populations are defined and who gets sampled as representative of that population. Which populations get selected to serve as representative genetic samples for each continent or region? Additionally, which populations get excluded to emphasize genetic difference between the continents? Considering the genetic diversity present in each continent, especially within Africa, what would it even mean to have "African ancestry" or "European ancestry?" Just like in admixture mapping, when ancestral populations are defined at the

---

[82] Shriver and Kittles, 207.

[83] TallBear, *Native American DNA*, 43.

continental level, the real genetic diversity present in each continent is erased. This also puts into the question the validity of the results, because if certain populations are excluded in an effort to heighten the differences between the continents, then the databases do not actually have an accurate representation of the various genetic markers that are present within the populations of a given continent.

A second major issue, which is the same as that faced by admixture mapping, is that while the tests claim to be looking at the DNA of ancestors, the tests compare a test takers' DNA to the DNA of various currently living populations. The AIMs used in these tests have been developed by comparing the frequencies of particular SNPs in the DNA of current living populations. The assumptions the tests rely on is that any markers that have been classified as AIMs found in a test taker's DNA must imply that they must have shared a common ancestor at some point with the current living populations who have that AIM. The results are at best guesses as to who one's ancestors might have been, not a definitive statement of who they were. This omission gives test takers a false sense of the accuracy about the results while freezing contemporary populations in time. It suggests that currently living West Africans, Koreans, and Western Europeans and so on have remained practically isolated and basically genetically the same for generations in order for their DNA to be seen as an adequate stand in for the DNA of contemporary American test takers' potential ancestors.

An issue that is relevant to both types of tests is the issue of transparency. DNA ancestry testing companies, for proprietary reason, are very tight lipped when it comes to sharing information about the DNA information they have in their databases and about how they go about calculating their test results. [84] This raises a number of concerns. One pertaining to the use of AIMs is, how exactly are the AIMs chosen? How much difference in frequency is needed to be considered an AIM and how reliable are they? While the Duffy Null allele is often used as a prime example of an AIM, it is rare that traits have such a drastic difference in frequency between populations like the Duffy Null. [85] The Duffy Null is unique because it is a trait that has an evolutionary advantage, which is why it makes sense it appears in higher frequency in an area where malaria is a concern. However, many of the AIMs that are used do not have such a clear division between populations, or only serve to distinguish between two populations but not between others. An example would be the marker known as SGC30055*1 which in the database used by the company DNAPrint occurred in 0.753 of Native Americans sampled and only in 0.054 of African Americans sampled and 0.511 of European Americans sampled. [86] This marker might be good for distinguishing between Native and African ancestry, but not between Native American and European Ancestry since there is a decent likelihood a test taker with that marker could have gotten it from their European ancestors. As TallBear points out, the markers can give us a hint, but they are not as clear cut as they seem at first. Without access to knowledge on how these companies are constructing and choosing their AIMs, there is not a clear sense of how accurate these estimates really are.

Another concern the lack of transparency poses is that test takers and other geneticists do not know the representation of each companies' database. Each company basically has their own

---

[84] Lee et. al., "The Illusive Gold Standard In Genetic Ancestry Testing," *Science* 325, no. 5936 (July 2009): 39,

[85] TallBear, *Native American DNA*, 72.

[86] TallBear, 73.

DNA database based on findings from past studies and the DNA that they collected from past test takers. The concern is that the databases could have an overrepresentation of samples from particular regions, such as Europe, and an underrepresentation of samples from other regions, such as Africa and Asia. This is in addition to the issue mentioned earlier of certain groups being selectively excluded from these databases. A potential consequence of this is that test takers with ancestry in regions that are underrepresented in a company's database can get very limited results that can be especially untrustworthy.

Bolnick et al. found that the selective exclusion of populations at the edge of various continents resulted in certain text takers receiving wildly inaccurate results, such as Middle Eastern, Indian, and Mediterranean test takers being told their DNA shows signs of Native American ancestry.[87] Euny Hong wrote of her experience taking a 23andMe test in *Quartz* where she got results suggesting she had ancestry from China, Japan, and only about 60% from Korea, even though both sides of her family are Korean.[88] After interviewing many geneticists and getting in contact with those working at 23andMe, she learned about the overrepresentation of European samples and relative underrepresentation of Asian and African samples in their database. In fact, 23andMe only had 67 Korean samples in their database of thousands. This lack of data results in the databases having a limited sense of the genetic diversity in particular regions, which leads to people getting test results suggesting that some of their traits most likely came from other populations because the statistical algorithms are making their best guesses as to where the test taker's genetic markers came from. The lack of transparency makes it hard to know how bad the lack of diversity in samples really is for each company. Additionally, the variance in the samples in each companies' database, as well as differences in how they decide to make guesses on one's ancestry means that one could get different results depending on what company one is using.[89]

The final limitation is that these tests cannot offer all the connections the test taker seeks. Those taking these tests are probably seeking information about their ancestors to learn more about their family history. This is especially appealing to those who do not have much information available to them about their ancestors and are seeking connection through these tests. However, the various limitations already addressed make it clear that these tests offer questionable results at best and unreliable information at worst. The other issue is that even if we take the test results at face value, they prioritize biological kinship over all other forms, which ignores the political and social elements of kinship and community. Being a member of a community is more than just being a biological descendant of one. Having a test result that claims one could have a "Native" or "West African" ancestor could provide interesting information, but that in itself doesn't mean members of that community or population would automatically accept the test taker as one of their own. Additionally, as the case TallBear provided earlier of the two indigenous grandparents illustrates, just because certain ancestries do not appear on a test that does not mean they are not a part of one's lineage. These tests cannot be the end-all and be-all of whether or not someone is a descendant of a particular community. Additionally, being genetically related to a community is not the only way to form connections. For example, there is a long history of Black people in the U.S. forming connections with

---

[89] Roberts, *Fatal Invention*, 248.

African communities in the African continent for various reasons, many of which have nothing to do with genetic kinship.[90] Genetic similarities is only one way of many for people to connect.

The various limitations of these tests put into questions the reliability of the results that they provide to test takers. However, DNA ancestry testing companies rarely present these limitations in their advertisements, and as a result they leave many of the assumptions about the relationships between race, ethnicity, nation, and genetics unchecked. Therefore, scholars have argued that DNA ancestry tests are contributing to a redefinition of race.

How DNA Tests are Redefining Race

The way the test results are given to consumers often conflates race, genetics, and ancestry. Companies like AncestryDNA and 23andMe who use AIM tests, which as mentioned previously often provide results using continental level descriptions of populations which mirror conventional race categories. The continental population categories of "Sub-Saharan African," "Native American," European," and "East Asian" can be easily interpreted by consumers as mapping onto familiar racial groups: Black, Native American, White, and Asian.[91] In practice, it appears that this is how many test takers are interpreting their results. For example, I have witnessed examples of test takers interpreting percentages of "African" ancestry as representing the percentage of "Black" ancestry in their DNA or interpreting percentages of "European" ancestry as representing their percentage of "White" ancestry. In practice then, one's genetics, hypothetical ancestry, and ideas about race get conflated and transformed into a one-to-one relationship. These tests suggest that one's ancestry is inscribed in their DNA and that a DNA test can accurately reveal to the consumer the various ancestral populations that shaped their

---

[90] Roberts, 253.

[91] Roberts, 228.

DNA. Consumers then interpret these estimates of ancestry as estimates of racial mixture. Each ancestral population becomes a conventional race category. These conflations are further reinforced by the narratives present in many ads that suggest a consumer has changed how they identified racially as a direct result of their genetic ancestry tests. Such is the case with an ancestry DNA ad that features a test taker named Livie, who prior to taking a test identified as "Hispanic" but after learning that she had ancestry from various places around the world felt compelled to start marking "other" on official documents that ask about her racial classification.[92]

When race, a social and political category, gets conflated with ancestry and genetics, troubling ideas about race get reinforced. Roberts argues these tests reinforce three central myths about race: that there are pure races, that each race contains people who are fundamentally the same to each other and different from others in other races, and that races can be determined biologically.[93] The idea that one's ancestry, which as I have discussed is often conflated with race, can be divided into segments that all add up to 100% seems to suggest the idea that pure "races" can exist or at least did exist in the past. To suggest one could be 20% European or 30% Sub-Sharan African implies that there is such a thing as a person who is 100% European or 100% Sub-Sharan African and that people can be classified as having "mixed" and "unmixed" ancestry. The continental level demarcation of populations in these tests flatten genetic diversity in the continents and support the idea that people of a particular "race" are more genetically similar to one another than to those of a different race when various studies have proven that

---

[92] Ancestry, "Livie: Father's Day," iSpot.tv, Accessed October 28, 2020, video, 0:29,

[93] Roberts, *Fatal Invention*, 228.

there is more genetic diversity among people classified as the same racial group. All of these

assumptions work to reinforce the idea that there is a genetic or biological component to race,

which is leading to what sociologist Michael Omi called the "(re)biologization of race."[94] Omi

argues that the expansion of the field of pharmacogenomics and the increasing use of DNA

testing for a range of purposes is leading to a reemergence of biologically and genetically based

categories of race.[95] The popularity of DNA tests means that the (re)biologization of race is not

limited to the labs of population geneticists and pharmaceutical companies, but is also subtly

being ingrained in how race is discussed publicly. More and more people are invoking DNA

when making racial and ethnic claims, and it seems as if people are starting to see genetics as

defining, or at the very least being a significant part of, identity.

The idea that genetic ancestry tests are enough to define one's racial and ethnic identity is

a form of "genetic determinism" that treats genes as the sole, or at least main, definer of one's

identity.[96] Once again, advertisements do much to support this trend by having testimonials

where people, as a result of their DNA test results, reject old identifications for new ones. Of

course, not all test takers will automatically take their test results at face value and change how

they identify accordingly. Sociologist Alondra Nelson has argued that people engage in

"affiliative self-fashioning" with their test results, meaning that they negotiate with their test

results, comparing it to other sources of knowledge they have about their past, and will accept or

---

[94] Omi, "'Slippin' into Darkness,'" 355.

[95] Omi, 351.

[96] Shriver and Kittles, "Genetic Ancestry," 209.

reject portions of the test results based on their own kinship and genealogical aspirations. [97] However, when the test results are taken as proof of identity in and of themselves, they can have negative consequences for racialized communities. Walajahi, Wilson, and Hull (2019) and Tallbear (2013) caution that the way Native American DNA is constructed in DNA ancestry testing company marketing perpetuates the idea that having genetic ancestry is all that it takes to be considered Native, and this could have the consequence of encouraging non-Natives to identify as Native on the basis of these tests alone.[98] There are already some who have used their test results as their main source of proof to gain access to tribal membership or resources set aside for indigenous people, such as special consideration in college admissions. [99] The use of DNA as the main, or only, source for claiming social and political identities are examples of Haraway's concept of "gene fetishism." [100] Haraway argues that key to gene fetishism is "forgetting that bodies are nodes in webs of integration."[101] By emphasizing the gene as the most valuable source of knowledge about the self, people forget that identities are constituted within social relationships. When DNA becomes the main source for making identity claims, the political histories of racial identities get de-emphasized and become much more individualized. For indigenous people, this can be a threat against tribal sovereignty, as native communities'

---

[97] Alondra Nelson, "Bio Science: Genetic Genealogy Testing and the Pursuit of African Ancestry," *Social Studies of Science* 38, no. 5 (October 2008): 763,

[98] Hina Walajahi, David R. Wilson, and Sara Chandros Hull, "Constructing Identities: The Implications of DTC Ancestry Testing for Tribal Communities," *Genetics in Medicine* 21, no. 8 (August 2019): 1748,

[99] TallBear, *Native American DNA*, 68.

[100] Haraway, *Modest_Witness@Second_Millenium,* 142.

[101] Haraway, 142.

right to define who is and isn't native is being claimed by DNA testing companies. Gene fetishism can be an issue for people of color more broadly, as it can lead to a de-politization of identities and promote racial identities based on a connection due to common biology as opposed "to common struggle for social justice."[102]

For this book, I am exploring themes in DNA ancestry testing advertisement and marketing to understand the narratives about race, ancestry, and community that are being produced in this marketing. One of my main research questions is: How do the histories of racial science and biological essentialism shape both the DNA ancestry test themselves and the claims being made by DNA ancestry testing companies in their marketing? The background provided in this chapter demonstrates the various ways the histories of racial science and biological essentialism are at play in the production and marketing of these tests.

In the next chapter, I conduct a textual analysis of ads that are focused on the theme of travel based on one's DNA ancestry. I apply Haraway (1997)'s concept of gene fetishism and Scodari's argument about the privileging of biological kinship to argue that these ads promote travel and connecting with people around the world on the basis of biological similarity. Additionally, the populations that travel's visit are represented in homogenous ways and shown for what they could offer the test taker on their journey of self-discovery. Genes become stand ins for sustained connections with others.

---

[102] Roberts, *Fatal Invention*, 257.

CHAPTER II: THE DNA JOURNEY—ANCESTRY TRAVEL IN DNA ADS

Introduction

In 2019, the *Today Show* did a feature where they followed Steve Balistreri, a resident of Naples, Florida who went on a trip to Sicily with his father to learn more about their family history.[103] The trip was a part of the new heritage tour packages offered by EF GO Ahead tours in partnership with Ancestry DNA. Prior to the trip, the Balistreri's were given DNA tests that showed them what regions of Italy their family most likely originated from, and they had the opportunity to meet with one of Ancestry's genealogists who was able to use their test results and Ancestry's database of historical records to tell them more about their family history.

Once in Italy, The Balistreri's were accompanied by another genealogist who takes them to the street where Steve's great grandfather used to live. While there they were able to visit the places in the neighborhood their ancestor may have frequented, including a local church where he and other ancestors most likely were baptized. The genealogist takes them to a local records office, which contained a book with the signatures of Steve's great grandfather and great grandmother before they traveled to the United States. They end their tour by having lunch in a restaurant owned by other Balistreri's who have organized a large feast that all members of the extended family could enjoy together. The feature ends with Keir Simmons, the main correspondent, stating "go out there and find out about your history before it's too late and now you have DNA to help you."

---

[103] EF Go Ahead Tours, "DNA travel: Go Ahead Tours on the Today Show," YouTube, February 22, 2019, video, 5:35,

The *Today Show* feature was tapping into what has become a major trend within the last couple of years: travel based on the results of one's DNA ancestry test. [104] In this chapter, I will briefly discuss the rise in interest in ancestry travel and the way DNA and travel companies have capitalized on this trend. Then, I will discuss how ancestry travel based on DNA tests compares to what is known about the practice of heritage or roots tourism before turning to an analysis of DNA ancestry testing ads that incorporate narratives and messages related to ancestry tourism.

The Rise in DNA Ancestry Tourism

The rise in popularity of DNA ancestry tests has been accompanied by a subsequent rise in people interested in taking heritage trips based on their DNA ancestry test results. Airbnb, in their market research, found that since 2014 the number of travelers who have used the service to trace their roots has increased by 500%, and prior to the onset of the COVID-19 pandemic *Luxury Travel Magazine* projected that ancestry travel would be a major travel trend in 2020.[105] Professionals within the tourism industry believe this boom in ancestry travel is part of a larger trend of travelers searching for personally meaningful and "transformational" travel experiences. By turning their DNA test results into travel plans it seems that travelers are seeking experiences that will stay with them long after the trip is over.

DNA testing and travel companies have sought partnerships that fulfil this increasing interest in ancestry travel. Ancestry DNA has engaged in many of these partnerships. The first partnership was with EF Go Ahead Tours in 2017 where they created the Heritage tours

---

[104] Melissa Wiley, "The Golden Age of at-home Genetic Testing May be Over — but it's Still a Critical Part of One of the Biggest Trends in How People will be Traveling in 2020," *Business Insider*, November 28, 2019,

[105] Wiley, "The Golden Age."

discussed above.[106] At the time, they created tour packages for Scotland, Italy, Germany, and Ireland, though as of 2021 they only seem to be running the Irish and German tours. In 2019, 23andMe partnered with Airbnb to offer services for clients interested in heritage travel trips. 23andMe would show test takers a collection of Airbnb rentals in the location that showed up on a test taker's DNA test; for example, someone with Mexican ancestry would be shown potential Airbnb rentals in Mexico City, while Airbnb would create dedicated pages corresponding to 23andMe genetic populations.[107] Many independent tour operators have also jumped on the rise in ancestry tourism, from smaller travel companies like Classic Journeys offering to match travelers with trips to places that show up on their DNA test and luxury companies like The Conte Club customizing trips that could span weeks.[108] It is clear from the uptick in these types of travel experiences that the tourism industry has taken note of traveler's desires for meaningful travel experiences.

Heritage/Roots Tourism

Heritage or Roots Tourism has historically been practiced by members of diasporas who either maintained a connection to the homeland or who longed to regain said connections.[109] Within the last few decades there has been an increase in people living in nations with long

---

[106] AncestryProGeneologists, "Genealogy Tours Led by Experts," Ancestry.com, Accessed December 21, 2020,

[107] Airbnb, "Heritage Travel on the Rise: Airbnb and 23andMe Team Up to Make it Even Easier," Airbnb News, Accessed December 21, 2020,

[108] Wiley, "The Golden Age"

[109] Dallen J. Timothy, "Genealogical Mobility: Tourism and the Search for a Personal Past," in *Geography and Genealogy: Location Personal Pasts,* ed. Dallen J. Timothy and Jeanne Kay Guelke (New York: Routledge 2016), 116-117.

histories of migration, such as the United States, Canada, and Australia, who have sought to learn more about their personal heritage by investigating and studying their family history and engaging in genealogical research.[110] This interest in one's past compels many to visit the lands of their ancestors to gain access to records, feel a deeper connection with their ancestors, and construct personal identities.

It is unclear what exactly spurred this interest in learning more about one's past, especially among people who are various generations removed from the immigration or diasporic experience. Timothy (2011) posits that the rapid pace of modernization and technological development in the global north might be producing anxieties about the future and as a result many people turn to an idealized version of the past to find grounding in an increasingly frantic world.[111] For members of marginalized racial or ethnic groups, the search for one's origins might be tied to feelings of unsettledness or rootlessness in the nation where they live. The search for one's roots is an attempt to build an identity that allows them to feel more at home.[112] Related to the latter, Nash (2002) suggests that genealogy is at once personal and political and that personal reasons for researching one's past often intersect with "wider cultural processes, politics, and social concerns."[113] Interest in Irish ancestry by white Americans can both be spurred by a personal interest in learning more about one's ancestors and prompted by discussions of

---

[110] Timothy, 118.

[111] Dallen J. Timothy, "Diaspora, Roots, and Personal Heritage Tourism," in *Cultural Heritage and Tourism: An Introduction* (Tonawanda, NY: Channel View Publications, 2011), 409.

[112] Timothy, 406.

[113] Catherine Nash, "Genealogical Identities," *Environment and Planning D: Society and Space* 20, no. 1 (2002): 30,

multiculturalism and anxieties about the increasing presence of Asian, African, and Latin American immigrants in the U.S. Nash also challenges Timothy's suggestion that the search of one's past necessarily results in a more grounded identity because while genealogy is often imagined as a way to find a stable identity, genealogy can also highlight the limits of historical knowledge and serve as a starting point for thinking of identity as "neither eternally fixed or essential, nor endlessly fluid."[114]

Timothy (2016) states that there are three anchors to roots tourism, the roots tourists themselves, the ancestral tour service providers, and government agencies that promote roots tourism.[115] The root tourists often engage in extensive genealogical research before traveling and are motivated to travel to gain access to new records, connect with ancestral lands, and build deeper connections with their ancestors and distant living relatives. Roots tourists often find such travel to be a spiritual or even religious experience, and it can be a cathartic experience for members of diasporas who were forcibly removed from their homelands. Tour service providers offer various services that might be of interest to roots tourists, including group tour packages, hometown visits, distant relative reconnections, and providing hands-on family history and genealogical research services. Finally, government tourism agencies in nations with long histories of emigration often actively promote genealogically based roots tourism. Examples include the Irish tourism board which has historically supported large gatherings of roots tourists and the Wales tourist board publishing guidebooks that assist roots tourists in engaging in family history in Wales.

---

[114] Nash, 49.

[115] Timothy, "Genealogical Mobility," 120-128.

The Genetic Ancestor

DNA ancestry testing companies have banked on this interest in roots tourism by framing their DNA tests as a new modern way to participate in genealogy for those interested in learning about their past. So called "ancestry travel" differs from older forms of roots tourism because DNA ancestry test replace extensive genealogical research and serve as the starting point for planning trips to one's ancestors' homeland. DNA ancestry testing companies appear to promote this form of travel as there are many ads that that either show people traveling to different locations based on their results of their DNA tests or testimonials of people who discuss their experiences traveling based on their test results. By conducting a textual analysis of these ancestry travel themed ads, I explore the ways that ancestry travel and ancestry travelers are shown in these ads and how the populations and regions the ancestry traveler visits are represented.

Matthew Elliot (2020) examined the reinvention of identity within DNA ancestry testing advertisements. In their analysis they discuss an ad from 23andMe titled "Reinventing Ancestry." The premise of the ad is that DNA testing companies like 23andMe are offering consumers a new way to approach ancestry and genealogical research. Instead of spending hours collecting family stories or searching through historical documents and public records, all the consumer has to do is send a sample of their DNA to a DNA testing company and through the science of genetics they will be able to learn about their ancestry through their 23 pairs of chromosomes. [116] Elliot argues that the reinvention offered by companies like 23andMe has no need for historical documents and family photographs because it does not need to identify

---

[116] 23andMe, "Reinventing Ancestry," iSpot.tv, Accessed December 21, 2020, video, 0:59,

specific ancestors. Instead, individual ancestors are replaced by the "genetic ancestor," a signifier created from genetic science that is "simply a conduit for global identity constructs, enabling consumers to make claims to certain populations and places."[117]

The use of the "genetic ancestor" is most present in ads showing test takers traveling or engaging in activities to learn more about their ancestral past. While the narratives of these ads often suggest that DNA tests can facilitate connections with people in different parts of the world and lead us to understand how we as humans are connected to one another, the appeal of the test conveyed through the visual elements of the ads are the cultures and places one would be able to claim through the genetic ancestors one can discover through a DNA test. The version of ancestry travel represented in these ads is centered on narratives of individuals engaging on journeys of self-discovery that incorporate people around the world on the basis of what they can offer the traveler. Since the genetic ancestor takes precedence over individual ancestors, people in other parts of the world are represented as members of homogeneous reference populations, such as "West African" or "East Asian." The specifics about what ethnic group within the broad category one's ancestors might have belonged to do not matter as much as the cultural capital of being able to claim that one is "East Asian" or "West African." The reinvention of ancestry promoted by DNA testing companies treats the gene as a replacement for extensive family history and genealogical research. Therefore, these companies are practicing what Donna Haraway termed "gene fetishism," treating the gene as the only thing needed to claim belonging

---

[117] Matthew Elliot, "'Amazing Stories Hidden Within': The Reinvention of Identity in Genetic Ancestry Advertisements," *The Journal of Popular Culture* 53, no. 1 (2020): 62,

and forgetting that bodies and identities are constituted within various social relationships, historical processes, and relations of power.[118]

I now turn to an analysis of the DNA testing company ads where ancestry travel is the focus of the ad. First, I examine how the ads represent ancestry travel as individual journeys of self-discovery, where the people one meets along the way exist on the basis of what they can offer the traveler. Then, I examine the ads where reference populations that test takers discover through their DNA tests are represented by images of scenic landscapes and architecture and the people in those locations are effectively erased.

Themes in DNA Ancestry Testing Advertisements

Individual Journeys of Self-Discovery

A central theme in these ads is the idea that the DNA tests, and the travel one may take in response to them, will allow the test taker to learn more about themselves and feel a greater connection to the rest of the world. The different populations that show up on a consumer's test are seen as representing different "parts" that come together to make up the person. To learn more about oneself, the test taker is compelled to take a journey of self-discovery to explore the places and cultures that show up on their DNA tests, and one of the ways to do this is by traveling to those locations.

An ad that perfectly illustrates this is a 23andMe commercial titled "Getting to Know You."[119] The title of the ad references the song with the same title from the 1951 musical *The King and I*, which plays in the background of the advertisement. The ad features Nicole, a light

---

[118] Haraway, *Modest_Witness@Second_Millenium,* 142.

[119] 23andMe, "Getting to Know You," iSpot.tv, Accessed December 21, 2020, video, 1:00,

skinned Black woman, as she travels to different parts of the world based on the results of her DNA test. In each location that Nicole visits, she is shown engaging in an activity with people local to the area, and these activities range from having a drink or playing chess to riding on bikes through sand dunes. After some of these activities, there is an accompanying graphic that shows up next to Nicole that displays a percentage that represents a result that Nicole got on her DNA test. For example, in one scene Nicole is shown under a dimly lit table with about three other people who appear to be Asian, and they appear to be in the middle of a toast. Immediately after the toast, Nicole is shown taking a selfie with the people she was toasting with and behind them there is a large crowd of people, some who are watching them and some who appear to just be dancing in the background. Therefore, Nicole is most likely right outside of a bar or a club. As Nicole is taking the selfie, the text "29% East Asian" appears right next to her on her left side and then a circle appears that encloses her head and, on the circle, there appears to be an orange line (pictured below).

Figure 1. Nicole taking a selfie at a bar in East Asia. (23andMe, "Getting to Know You.")

Based on what is shown in the rest of the ad, it appears the circle that appears above Nicole's head is meant to represent her DNA, or at the very least her DNA as represented by her 23andMe test results, because the circle is always accompanied by a text that references a percentage and a 23andMe reference population. Additionally, the length of the orange line is meant to represent the proportion of Nicole's DNA that is "East Asian" since throughout the ad the lines are shorter when the percentages are smaller and longer when the percentages are larger.

Returning to the scene at the bar, excluding the graphic that reads "29% East Asian" there does not appear to be any information as to where Nicole is located. As a result, the bar, and the people Nicole is surrounded by, become a stand in for the entire region of "East Asia." The same thing happens for all the locations that Nicole visits throughout the ad. There is a scene where Nicole is getting a massage in what seems to be a spa where one is only able to see the masseur and some people in the background for a split second before getting a close up on Nicole that is accompanied by a graphic that reads "12% Middle Eastern." This is followed by a scene of Nicole joining in on a game of soccer and then dancing with a group of people down a road in what appears to be a semi-rural area accompanied by the text "46% West African." In each of these scenes the viewer is not given much information about the specific locations Nicole is visiting. Unless someone is already familiar with the locations where the ad was filmed, the only cues a viewer is given as to where Nicole may be are the people she interacts with and the scenery of each location. As a result, the people and the scenery at each location symbolically stand in for larger geographic areas. The Asian people Nicole hangs out with at the bar are stand in for all of "East Asia." The Black people Nicole plays soccer with in the semi-rural town stand in for all of "West Africa." The lack of specificity works to homogenize these populations as audiences are not given information about what communities in East Asia or West Africa Nicole's ancestors may be from.

At the very end of the ad, there is an image of Nicole's face with a circle made up of various colors, each representing the populations that she visited and that showed up as results on her DNA test, and the text "100% Nicole." Throughout the ad, there were phrases that pop up in between the scenes of Nicole visiting the various locations on her journey. When put together the phrases read "there are parts of all of us yet to be discovered and through our DNA we are all connected." The ad ends with a voice over that represents the company 23andMe that says, "learn how your DNA connects you to the world; know more about you and the 23 chromosomes that make up your DNA at 23andme.com" over the phrase "Celebrate your DNA." Through phrases such as "learn how your DNA connects you to the world" and "through our DNA we are all connected," 23andMe promotes the idea that a DNA test can show an individual test taker how they are a part of a wider global community and that DNA tests can show us all that we are more connected with each other than we often realize. The ad suggests that DNA tests will allow test takers to both learn about themselves and make connections with other people around the world.

The premise of the "Getting to Know You" ad is that Nicole is going on a journey to all the places that appeared on her DNA test to visit all the places her ancestors may have originated from to learn more about herself. However, the "you" that Nicole is getting to know is not the people she visits; it is herself. The viewer gets little in the way of information about the people Nicole visits, and the people she encounters on her journey exist to provide a service or show her a good time, such as the people in West Africa who played soccer and dance and the street with her or the women in the Middle East who serves as her masseur. So little is known about them that the audience does not even know what specific nations they are in. All they get is that Nicole is somewhere in "East Asia," "West Africa," or "The Middle East." The most specific region identified is "Scandinavian" and even that is still a term referencing a subsection of Europe that includes multiple nations and communities. The people she visits become representatives of geographic regions where Nicole's ancestors may have descended from. As Elliot (2020) articulated, individual ancestors with specific histories are erased and replaced with "genetic ancestors" whose value comes from being able to make claims to regions of the world. The appeal is being able to claim "East Asian" or "West African" ancestry not necessarily learning about the communities within these areas that specific ancestors are from. The vagueness of the labels for the reference populations reflects the homogenization of these large populations that occurs in the DNA tests themselves. While DNA companies like 23andMe suggest that DNA tests can lead us to understand how we are all connected, the people test takers are meant to connect with are homogenized in the process.

Meaningful Travel

Another reoccurring theme in these ads is that ancestry-based travel is different from traditional tourism because it is more personally meaningful since one can claim to visit these places because it is a part of their heritage as opposed to simply being a tourist. In the Ancestry ad titled "Discover your heritage," a woman is shown in her living room looking at the results of here DNA ancestry test. She discovers through the test that some of her genetic ancestry comes from Italy, specifically Sicily. [120] In the next scene, the woman is transported to Italy where she is shown asking a vendor for directions. She walks up to a small house and right before she knocks on the door, we see on her phone an Ancestry profile for of a man who Ancestry shows as being one of the woman's third cousins. Through this reveal, we learn that the woman came to Italy to

---

[120] Ancestry, "Discover Your Heritage," iSpot.tv, Accessed December 21, 2020, video, 0:30,

visit her extended family in Italy. After she introduces herself to her newly discovered third cousin, there is a shot of her eating a meal with her cousin and his family in a grassy backyard. There is then a transition from the family meal to the woman back in her apartment talking with two other women who appear to be her friends. She holds ups a wine bottle which has the same label as the bottle that was present at the family dinner she had back in Italy. There is voice-over present throughout the ad that says that a DNA test could "lead you on an unexpected journey that leads you closer to home" and that "instead of telling stories of where you went, you can tell the story of where you come from."

The last quote is said over the part where the woman is holding the bottle of wine and talking to her friends. The line suggests that the travel that one engages in based on their DNA test results is significantly different from traditional travel or tourism because you "come" from those places and that by visiting them you are getting "closer to home" and therefore you have a more personal connection to them. The woman in turn can tell a much more profound story, one about her family history and the family she was able to reconnect with instead of simply a story of the places she went to on vacation.

The idea that this ancestry-based travel is different from traditional tourism is reinforced symbolically through the recurring narrative of family reunification. The inclusion of distant relatives reinforces the idea that you are traveling not necessarily as a tourist, but as a relative who has not "come home" in a while. The people one meets on the journey can be both literally family, as in the case of meeting distant relative for the first time, and figuratively family in the sense that the traveler is visiting the place where their ancestors are from and the people in the region could be considered to be distant relatives. It also symbolizes the idea that through DNA we learn that we are all a part of a global human family. However, what does it mean to be "from" somewhere that one did not even know about until they got a DNA tests result? These ads suggest that all one needs to claim belonging to a place or a community is that it shows up on a DNA test.

Belonging through Consumption

Another idea present in these ads is that DNA tests provide you with cultures and places to explore, and much of this exploration is based in consumption. An ad that exemplifies this is an Ancestry ad titled "A world of new cultures."[121] The ad is structured in a similar way to the "Getting to Know You" ad; it features three different DNA test takers who engage in various activities or visit places representative of the reference populations that showed up on their DNA test. Like in the other ad, there is a graphic that appears next to the test taker with a percentage and a reference population. The first person shown is a man sitting in a restaurant looking at his phone while a server is coming into the frame to bring him his food. On the phone, there is a screen with the man's Ancestry DNA test result page and one can see a map of Europe where a red circle is placed on Spain, implying that he got DNA test results that suggested he had "Spanish" Ancestry. Once the phone is lifted, one can see the plate of food that the server brought and it looks to be a plate of paella, a well-recognized dish from Spain. The camera then shows the server talking to the man at the table, presumably about the food he just brought, and next to the man appears a graphic with the text "38% Spanish" ancestry (pictured below). The man is then shown conversing with the waiter using a Spanish-English dictionary.

---

[121] Ancestry, "A World of New Cultures," iSpot.tv, Accessed December 21, 2020, video, 0:30,

Figure 2. Man ordering at a Spanish restaurant. (Ancestry, "A World of New Cultures.")

In the next scene, there is a woman walking by a fruit stand who stops and shows the fruit stand owner something she has on her notepad. As she is showing the man her notepad, a graphic that reads "29% Polynesian" appears next to the woman. The man then brings a fruit, which suggests what she had written was most likely the name of a specific fruit she was looking for. What stands out about this scene is that while in the prior one it could be assumed the man was in a Spanish restaurant, there are various ways this scene could be interpreted. The woman could be in a food market somewhere in Polynesia, the fruit stand could be owned by someone who is Polynesian, or the woman was looking for a fruit that originates from Polynesia. There is not enough context to know for sure. The last scene shows a younger and older woman in a kitchen who are cooking something from a recipe in a cookbook. When the younger woman goes to check the recipe from the book, one can see the title of the book: *Traditional Lithuanian*. As the women are talking, there is a graphic that appears next to the younger woman with the text "25% Baltic." The ad is accompanied by a voiceover that runs through the whole ad and states, "Ancestry DNA can open you to a world of new cultures to explore" and "You can connect more deeply with the places of your past and be inspired to learn about the people and traditions that make you, you."

The phrase "Ancestry DNA can open you to a world of new cultures to explore"
explicitly suggests that an appeal of these tests is that they will inform you about heritages you
may have not been aware of and you will have new cultures to both claim and explore. The type
of exploration of cultures shown through the ad is based on a form of consumption, in this case
specifically food consumption as seen by the fact that all three are engaging in activities that
involves food (eating at a restaurant, finding food at a market, cooking food from a cookbook).
Additionally, two of the three people who are not test takers who are shown in the ad are in
positions of service to the test taker exploring new cultures, one as a server in a restaurant and
the other as the food stand worker. Even at the end, the voiceover explicitly states that one can
"savor your DNA story" by using one of Ancestry's DNA tests. A focus on food consumption
reinforces the idea that the travel is for the individual person doing the travel as they are not
necessarily connecting with the people they meet through their consumption practices, especially
since they people they engage with in these practices are often service workers. The two parties
in this interaction are then not connecting at an equal position. The residents in these destinations
end up being background characters in the traveler's journey of self-discovery, and they are often
workers in the service economy.

Something else to note about the ad is that the homogenization present in the "Getting to
Know You" ad is also present in the case of the women with the graphic that read "29%
Polynesian." In an ad where the focus is about exploring new cultures based on results of a
DNA test, what would it mean to "explore" Polynesia when the region known as Polynesia
encompasses various ethnic groups and nations. Without more specific information about the
people, one's ancestor may have descended from, all a test taker could do is explore cultural

practices from various ethnic groups within Polynesia, which ends up homogenizing these communities and treating them as interchangeable.

While at one level the ads suggest that DNA tests can be used to facilitate connections with people around the world and learn how one is part of a larger global community, that connection relies on what others can offer the test taker on their journey of self-discovery. The act of homogenizing the communities which the test takers meet shows a lack of care when it comes to learning about people around the world since the test taker is not taking the time to learn more specifically which communities their ancestors may have descended from. Instead, the people they come across in their journey have value in what they can offer the traveler, be that a paid service or as representatives of cultures the test taker wishes to claim. Instead of promoting great connection with others, this promotes a greater misunderstanding of the various communities that exist in the world.

Traveling for the Places or the People?

There is a collection of ads that, when discussing the experience of a test taker traveling to a location to visit distant relatives they had recently learned about, tend to emphasize the scenic and historic locations their distant relatives took them to visit. The Ancestry ad "Michael" features a testimonial from an Ancestry DNA test taker who learned about a distant cousin in Ireland through Ancestry DNA's database. [122] After connecting with said cousin through Ancestry DNA, he planned a trip to Ireland where he met "another twenty cousins" who took him to various locations including "the Cliffs of Moher, their ancestral home, and the family bar." As Michael is talking about the places they visited, there are moving images of buildings

---

[122] Ancestry, "Michael," iSpot.tv, Accessed December 21, 2020, video, 0:30,

and landscapes representing the places he visited. There is an image of the Cliffs of Moher when he mentions them, there is an image of a pink flower in front of a stone building when he mentions "the ancestral home" and a shot of a camera looking at a window from within a dimly lit building accompanied by the sound of glasses clanking against one another when he mentions "the family bar." Michael states that the journey he took because of his DNA test gave him "a sense of connection to something bigger than yourself." At the very end of the ad there is a shot of Michael holding out his phone to the audience; the phone screen shows a highlighted map of Ireland. Underneath the map are what appear to be a group of profile pictures, which most likely represent the cousins that showed up in Ancestry's database. Under the profile pictures is a collection of landscape photos meant to represent various locations within Ireland (pictured below).

Figure 3. Michael showing his ancestry test results. (Ancestry, "Michael.")

What stands out from this ad is that while Michael's narration focuses on the familial connections he made; the imagery is focused on the scenic locations he visited. At no point in the ad are there images of people other than Michael and even at the end what takes up most of the screen are the images of the locations he visited not the cousins he met. Michael states that learning about his cousins and going to meet them in Ireland gave him a sense of connection to something bigger than himself, which suggests the idea of feeling more connected to the rest of the world. However, that is not what is communicated visually. Instead, it seems what is being marketed are the scenic places one would be able to visit and claim as a part of one's heritage because of their DNA and not necessarily the connections with people across the world that the DNA test could facilitate.

The focus on locations over people is also present in ads that feature testimonials of people who have not engaged in ancestry-based travel and are simply recounting the discoveries they made through their DNA tests. Examples of these ads include the Ancestry ads "Krystina" and "Ruben." In both ads, the person giving the testimonial had recently taken an Ancestry DNA test and they talk about their experiences receiving their results.

Krystina had taken the test to learn more about her heritage and was shocked by what she discovered.[123] She learned that she had ancestry from Cameroon, Congo, the Bantu People, the Ivory Coast, Ghana, and Togo. As she discusses her results, the advertisement features moving images that appear to represent the various populations she lists since a new image appears after she mentions a new name. Like in the Michael ad, the most of what is shown are still and moving images of landscapes and an aerial shot of a town. People are shown for about 1 to 2 seconds in total of the 0:30 minute ad and each time they are only on screen for less than a second. There is a quick shot of three small children playing soccer, an image of someone's hand pulling on clothes on a wire, and an image of a woman walking with a basket of what appears to be wood as she faces away from the camera. One cannot see the faces of anyone in any of these

---

[123] Ancestry, "Krystina," iSpot.tv, Accessed December 21, 2020, video, 0:30,

images. In effect, the people that exist in the locations mentioned by Krystina are erased, as if all that exists in those locations are scenic environments and buildings.

In the Ruben ad, Ruben had recently taken a DNA test and said that the test "saw" and "recognized" the regions where his family was from, which appears to imply he already knew of his family heritage and the test was just confirming what he knew as opposed to giving him new information.[124] The test results stated that that his family was from the State of Jalisco, Mexico and especially the city of Guadalajara. The images that were shown as Ruben was speaking were either quick glimpses or longer aerial shots of the locations he references, such as an aerial image of Guadalajara when he mentions the city by name. When it comes to images of people, in the ad there was a quick shot of the legs of someone working in a field, two children in a store with one facing away from the camera, and someone's hand flipping a small tortilla on a stove. The only other images including people were a collection of black and white historical sketches from what might be the colonial era. Both Ruben and Krystina's ads end in the same way as Michael's did, with them pointing their phone screens to the camera showing a highlighted map which represents their DNA ancestry test results. However, unlike Michael, there are no images of potential cousins under the map. Instead, there are smaller versions of the images of landscapes and aerial shots of cities used in the ad (pictured below).

---

[124] Ancestry, "Ruben," iSpot.tv, Accessed December 21, 2020, video, 0:30,

Figure 4. Krystina showing her ancestry test results. (Ancestry, "Krystina.")

Figure 5. Ruben showing his ancestry test results. (Ancestry, "Ruben.")

What stands out about these ads is that the test takers did not travel to the places that showed up as results in their DNA tests. Instead, the narrative focuses on how these test results allowed them to feel proud about their family heritage and more connected to who they are. Yet the visual focus of the ads are aerial shots of the cities or nations mentioned, with the people of those places effectively absent from the landscape. These places are represented as a collection of landscapes and architecture and not a collection of people. This is further reinforced with the last image of each ad having the DNA results represented by images of a highlighted map above a collection of landscape and architecture images leaping off the phone. It seems then that there is a disconnect between what the narrative of the ads verbally suggest is the appeal of these tests (i.e., learning about family history, meeting and reconnecting with distant family, feeling like one is a part of a larger collective) and what is visually represented in the ads. The images suggest that the draw for the test taker is not necessarily the people they would be able to meet through these tests, but instead the places and cultures they would be able to claim as a part of their heritage.

Conclusion

DNA Ancestry testing companies take advantage of the increased interest in ancestry travel by promoting the idea that their DNA test can offer test takers a new and modern way to engage in genealogy. Instead of extensive genealogical or family history research, through a simple DNA test one can learn about where one's ancestors came from and possibly find distant relatives. These ads claim that if a test taker does not have information about specific ancestors, they are able to make a claim to belonging to a place and its people through a "genetic ancestor" provided by these tests. Test takers will be able to learn more about themselves, while at the same time building connections with people around the world. DNA and DNA tests supposedly can bring us to an era of greater human connection.

However, the centering of individual journeys of self-discovery and individual consumption practices represents the test taker making new connections on the basis of what others can offer them on their journey. The people the test taker encounters are often represented without specific reference to the communities they belong to and are instead framed as representatives of broad regional labels. They exist to provide services to the test taker and their value comes from being representatives of the communities and cultures the test taker wishes to claim. The framing of these ads also come with some colonial implications. One of the major assumptions in DNA ancestry tests is that contemporary populations, whose DNA is used to make the reference populations in the test, are genetically similar to the populations who lived in those regions hundreds of years ago and therefore can be used to make calculated guesses of the genetic ancestry of test takers in the global north, particularly the U.S. and Europe. These populations are also assumed to be culturally similar to these ancestral populations are test takers take these trips in order to learn about their past and reconnect with their roots by visiting these populations. These populations are then relegated as people of the past both genetically and culturally. This neglects that fact that over many years populations migrate and intermix so one cannot assume that contemporary populations in Africa, Asia, and the Americas are the same as the populations who lived in those regions hundreds of years ago.

Another colonial aspect of these travel narratives is that these communities do not get much of a say in whether the traveler is able to claim to be a member of their community. The ads suggest that all that is needed to claim belonging is that a particular region or community shows up on one's DNA test. It treats belonging as simply a matter of genetic descent, completely ignoring all the social and political dynamics that go into claiming being a member of a certain community. How can one claim to be West African, Scandinavian, or East Asian if they do not even have some knowledge about specific communities within those categories that one's ancestors may have been from? Additionally, is claiming communities simply on the basis of a DNA test not taking away the agency of communities to decide who is and is not a member of their community? Said connection is based on the terms of the traveler alone.

The way communities around the world are represented in both the ads and the tests themselves promotes less, not more, understanding as diverse heterogenous regions are represented as homogenous and interchangeable. This is especially the case for communities in the global south as they are often represented using vary broad categories, such as West African,

East Asian, Middle Eastern, while populations in Europe tend to be represented as a smaller scale, such as test results that specify "Italy" or "Scandinavia."

As Nash (2002) argues, genealogy has the potential for critical engagement with one's identity. If test takers take these DNA tests as a starting point for further family history and genealogical research, as well as fully understand the limitations of the test results, there is a potential for the test to promote an understanding of identity and ancestry as complex and maybe provide test takers with the connections and meaningful experiences they seek. However, this is not what is depicted in these advertisements. What is depicted instead is travel that reflects a very limited understanding of one's connection to the world and that connection is based on biological similarity and individual consumption that further erases and homogenizes the people with whom one is supposed to be connecting.

In the next chapter, I will turn to the themes of DNA tests serving as tools to fight against racism. It seems that there is a paradox that the tests purport to challenge racism by proving it is illogical while the test themselves are based on racist principles stemming from old ideas of racial science. I will outline how race and racism are being defined through these advertisements to get a sense of how they are settling this internal contradiction.

CHAPTER III: RACE AND RACISM IN AN ERA OF DNA TESTS

Introduction

DNA ancestry ads often conflate race, ethnicity, and nationality and reinforce messy relationships between the three and genetics. An ad where this is prominent is the Ancestry DNA Ad titled "Livie."[125] The beginning of the ad shows an older brown skinned woman sitting on a couch, with two floral print pillows at her side, in front of a white wall. Livie begins her testimonial by stating that in the various countries to which she has traveled, people would ask her "What is your nationality?" and that she always answered by saying she was "Hispanic." She goes on to talk about how when she got her DNA results, she was shocked because she was "from all nations." As she says this, a multicolored pie chart appears next to her showing the various reference populations that she received in her Ancestry DNA test results. As seen in the figure below, she has two large percentages, represented by Native American (33%) and Spain/Portugal (31%), and the remaining 36% of the pie chart is composed of many smaller percentages, including Ireland, Italy/Greece, Middle Eastern, and African.

---

[125] Ancestry, "Livie: Father's Day," iSpot.tv, Accessed February 19, 2021, video, 0:29,

Figure 6. Livie and her ancestry test result pie chart. (Ancestry, "Livie: Father's Day.")

These results seem to have posed a new problem for Livie, as she claims that after

receiving her results, she would look at official forms and not know "what to mark" because

"she's everything." She resolves this new dilemma by stating she now marks the "other" box.

This last detail suggests that Livie is making reference to documents that asks demographic

questions about a person's race and ethnicity, as "other" is often a box one is able to pick on such

forms and because the trope of not knowing which box to pick is often associated with

multiracial people being unsure of which racial "boxes" to check on official documents or

arguing that only having to pick a single category does not adequately represent their racial

identity.

Throughout this ad, it is evident that the categories of race, ethnicity, and nationality are

conflated with one another. From the start, Livie states that when others would ask what her

nationality is, she would answer that she is "Hispanic," which is not a nationality but a pan-

ethnic category that could be applied to people of various nationalities. She then argues that her

DNA test results, which include a variety of reference populations representing particular

nationalities (Ireland), geographic regions (African, Middle Eastern), and a racial/political category (Native American), as proof that she is "from all nations." This information then shapes how she identifies racially and ethnically by claiming that on official documents she now marks "other" as she believes none of the other boxes adequately represent her ancestral origins. The lines between ethnicity, race, and nationality become blurred. Additionally, the shift in identification based on the results of a DNA test treats racial, ethnic, and national identities as a matter of genetic inheritance. The narrative of the ad implies that she changed how she identified racially and ethnically solely because of the results of her DNA test. There is no mention of how the results relate to her knowledge about her family history, or the social and political element of racial and ethnic labels.

The conflation of race, ethnicity, and nationality alongside the geneticization of these categories allow for DNA ancestry testing companies to make another interesting claim: that DNA tests can be used to fight against racism and xenophobia. A narrative in some ads suggests that DNA ancestry tests can show us a different version of human history rooted in genetics that shows us that over time humans have intermixed with each other so much that there are no clear genetic divisions along racial, ethnic, or national lines. They suggest that the labels we use to classify one another do not adequately represent the various lineages we each have and that the idea of "pure" races or ethnicities does not exist. The thinking goes that if more people realized that we have more in common than not, then people would be less inclined to believe and perpetuate racism and xenophobia. However, for this logic to work, race, ethnicity and nationality need to be understood as labels used to describe a person's ancestry. In the process, racial, ethnic, and national categories are treated as matters of genetics as opposed to social and political categories emerging out of the history of colonialism and the social structure of a given society.

As race, ethnicity, and nationality are treated as personal and de-politicized identities, racism and xenophobia also must be individualized and de-politicized. In this chapter I will be making a distinction between prejudice and racism. I borrow Bonilla-Silva's (1997) definition of prejudice as negative attitudes held by individuals about others who belong to certain socially defined groups that can lead individuals to act in discriminatory ways towards those groups. I borrow from Mize (2013) to define racism as the structure of a society where social institutions, interpersonal interactions, and ideologies work in conjunction to unequally distribute economic, political, and social resources and rewards along racial lines.[126] I understand prejudice as one of many elements of systems of oppression like racism that helps maintain the status quo. In these advertisements, racism and xenophobia are portrayed as being mostly the result of prejudiced individuals, as opposed to being systems built into the structure of a given society that unequally distributes social, political, and economic resources and rewards to benefit groups in positions of power. This reasoning not only leads to a depoliticization of race and ethnicity, but it also helps DNA testing companies to claim that their test can hold an important space in the public discourse concerning antiracism and social justice while simultaneously hiding the scientific practices these ancestry testing companies engage with which rely on and reproduce racist assumptions about human genetic diversity in the first place, such as, the assumption that those in the same racial groups are more genetically similar to one another than to those in another racial group. The result is a misunderstanding of racism and a perpetuation of biologically essentialist notions of race, ethnicity, and nationality.

---

[126] Eduardo Bonilla-Silva, "Rethinking Racism: Toward a Structural Interpretation." *American Sociological Review* 62, no. 3, (June 1997): 466, Mize, "Critically Interrogating the Illusive Sign," 359.

I will begin by expanding on how race, ethnicity, and nationality, and the relationship between the three, are represented in these ads. Using a few examples, I will show how racial and ethnic identities are boiled down to one's "genetic mix" and how the ads perpetuate the idea that one "inherits" certain identities regardless of one's personal experiences. This relates to another pattern found in the ads: removing "race" from its social and political context and treating it simply as a descriptor of one's ethnic background (i.e., to be "white" is to have European ancestries and to be "Black" is to have African ancestries). I will then provide analysis for ads that suggest that DNA tests can be used as tools against racism and xenophobia, showing how racism or extremism is being defined in that context. I will end by thinking through the paradox of DNA companies claiming to be tools of antiracism while the test themselves rely on assumptions about human genetic diversity rooted in a history of racial science.

Identity Based on One's Genetic Mix

The focus of this section is exploring how the concepts of race, ethnicity, and nationality are conflated in these advertisements and how the ads remove the social and political context of these categories and treat identity as solely derived from one's genetic mix. These themes are central to the narrative of the Livie advertisement discussed in the introduction. Racial, ethnic, and national categories are treated as if they are interchangeable and Livie changes how she identifies based on the results of her DNA test. Marking "other" on forms implies that DNA test should be treated as a source of knowledge on one's true ancestral background and that racial, ethnic, and national labels are simply descriptors of one's ancestry and mostly determined by a person's genetics. Racial, ethnic, and national categories are depoliticized and separated from histories of racism and colonialism.

Another advertisement that reinforces these themes is the MyHeritage DNA ad titled "Humanity."[127] The ad features a spoken word poem performed by an artist named Prince Ea. For this section I will focus on the first half of this ad as it contains many of the conflations between race, ethnicity, and nationality present in the Livie ad. Prince Ea begins standing in the center of room with white walls. He directly speaks to the audience and says, "I don't know about you, but when I fill out applications, I never know which race box to check." After he says this, the words begin to appear on the wall around him. The first five to appear are "Native American," "Hispanic," "Asian," "White," and "Middle Eastern." This question is meant to seem non-sensical to the audience as most would identify Prince Ea as a Black man and would think the answer to the question would be obvious. Prince Ea acknowledges the audience's confusion by saying "you are probably thinking to yourself, 'it's obvious, your Black.'" This is a set up for him to offer a rebuttal by discussing his results from his MyHeritage DNA test. He says that the test showed him that his DNA is actually, "72% West African, 14% British, 7% West European, 3% East European, and 3% Finnish." As he is describing the results from his DNA test, a map appears on the wall behind him, and different regions get highlighted as he mentions them and their corresponding percentage on his DNA test. After listing off all of his results he talks to the audience again and says, "now you tell me, which race box do I pick? and If I pick that one, do the others get dismissed?"

Like the Livie ad, the focus is on how one's DNA test results have reconfigured how they identify, especially when it comes to being asked their race and ethnicity on official documents. He appears to have the same dilemma Livie does: his DNA test results have given him a new

---

[127] MyHeritage DNA, "Humanity," iSpot.tv, Accessed February 20, 2021, video, 1:00,

understanding of his ancestry and of himself so now he does not know which race box to pick because he believes none of them fully encompass his ancestry. However, the reference populations in his DNA test, which are the evidence that prompted this dilemma, are not racial categories. He says that his DNA test results showed that he has genetic ancestry from Western Africa, Britain, Finland, Western Europe, and Eastern Europe. These reference populations stand in for either broad geographic regions or specific nations, but they are not "races" in the way race is understood in the U.S. However, the premise of the ad suggests that it was these results that led Prince Ea to not know which race box to pick. That either means that the reference populations given in the ad are being treated as if they are different "races" (i.e., West African and Britain are treated as "races") or that the results are being interpreted in a way that associates particular racial categories with the part of the world the reference population is located in (i.e. West African ancestry is associated with the racial category "Black" while West European ancestry is associated with the category "White"). Either way, the result is that Prince Ea now defines himself as multi-racial and believes that having to only choose one "box" on forms that ask for his race ignores the fact that because of his ancestry he belongs to various races.

The shift in Prince Ea's identification being shown to be based solely on the results of his DNA tests highlights the fact that "race" in these ads is represented as being mostly, if not solely, about biological ancestry. Race is boiled down to genetic ancestry, where particular geographic regions are associated with certain racial categories and inheriting DNA from populations in certain regions is enough to claim belonging to various racial categories. This erases the fact that racial identities are political identities emerging out of histories of racism and colonialism. They are not simply reflections on one's family background but products of the history of groups in power racializing and defining others as belonging to certain racial categories in an effort to deny them access to power and resources. Over time, those who were racialized and subordinated have struggled to dismantle the unequal distribution of power and resources and have worked to reconfigure the meanings associated with the racial categories that have been imposed on them. Claiming particular racial identities, such as Black, is as much about describing one's social position in society and claiming affiliation with those who are similarly racialized as it is about describing one's ancestry. Boiling race down to genetic ancestry completly negates this political dimension.

Equating identity and belonging to genetic inheritance is present in ads where people change their ethnic identification as a result of DNA test such as Ancestry DNA's ad titled "Kyle."[128] Kyle begins his testimonial by stating that growing up his family identified ethnically as German. He also participated in cultural practices associated with Germany, such as dancing in a German dance group and wearing lederhosen. This set up explains his surprise when he first started using Ancestry's family tree service and finding that he was not seeing a lot of Germans in his family tree. As a result, he decided to take a DNA test and through the test he discovered that they were not German at all, by which he means that Germany did not show up in his DNA test. Additionally, he discovered that 52% of his DNA was classified as "Ireland, Scotland, and Wales." This result led Kyle to trade his lederhosen for a kilt, symbolically shedding his former German identity for a Scottish and Irish one.

What stands out about this example is that Kyle's prior ethnic identity, which is based on his childhood experiences and immediate family history, is discarded on the basis of an ancestry test. The discrepancy that Kyle observed in his family tree, that there were not many Germans,

---

[128] Ancestry, "Kyle," iSpot.tv, Accessed February 20, 2021, video, 0:30,

could have an explanation that does not negate how his family identified growing up, such as that a couple generations back a non-German ancestor may have moved to Germany and the subsequent generations started identifying as Germans. Instead, Kyle decides to turn to a DNA test to find answers. Kyle could have received his DNA test results and still maintained the ethnic identity he was familiar with, but instead he chooses to refashion his identity based on the results of his DNA test without much hesitation. The shedding of an ethnic identity based on familial upbringing in favor of one based on a genetic ancestry estimate removes the social element of ethnicity. As with the examples of Prince Ea and Livie, identity is removed from social and political context and based heavily on the results of a DNA test.

This points to an underlying theme in each of these ads: advances made by new technology are allowing us to learn new and more "accurate" information about ourselves and our ancestry. In each ad, the protagonist treats how they used to identify as misguided and based on either a false or limited understanding of their family's past. The most Livie knew about her nationality is that she is "Hispanic," Prince Ea used to only identify as Black, and Kyle used to identify as German since that is how his family identified growing up. All three were amazed by their results in their DNA tests and ended up changing how they identified without much question. The test results are treated as reliable sources of knowledge about oneself, more reliable than other types of knowledge such as family narratives or personal experience. If family history and experience were treated as equally important and a reliable source to base one's identity, they all would not be so willing to change how the identify. Within these ads there is also no discussion of what is lost in this reshaping of their identities. Returning to the Kyle advertisement, his old German identity was rooted in the stories he was told by his family and the cultural practices in which they engaged. There is no mention of what it would mean to let go of an ethnic identity that is deeply ingrained in one's family for one that they only recently learned about because of the DNA test. Instead, Kyle appears to replace his lederhosen for a kilt without much thought. These ads then imply that you might think you know yourself and your family history, but that a DNA test is the only "true" way to know since it can give you a more scientific way of knowing about your past. As these results are supposedly more reliable, how you identify should be based on the results of a DNA test and one should be willing to shed how they previously identified. Racial, ethnic, and national identities become nothing more than labels that describe one's genetic mix.

DNA Tests as Anti-Racist Tools

An ad where this theme is significant is titled "Momondo: The DNA Journey."[129] Momondo, a travel deals website, collaborated with Ancestry to produce an ad that shows the world "there are more things uniting us than dividing us." The premise of the ad is that a random chosen group of people from across the world are taken on a "DNA journey" where they take a DNA test to learn more about their ancestry. The first part of the ad begins with the question "Would you dare to question who you really are?" written in white font on top of a black background. This is followed by segments of interviews with the participants which appear to have occurred prior to them receiving DNA results. In these clips, participants discuss their family background and how they identify. Most of the respondents define themselves by the patriotism they have for their country, such as a man who discusses how his family has served in the British military and another who claims he is very patriotic about Bangladesh. One of the respondents defines themselves racially, describing her family as "just proud blacks," and

---

[129] Ancestry, "Momondo: The DNA Journey," YouTube, Accessed February 21, 2021, video, 5:16,

another woman shows the interviewers a picture of her mother in traditional Kurdish clothes to signal that she mainly identifies as Kurdish, which is an ethnic group.

This segment is followed by one of the interviewers posing the question: "Think about other countries and other nationalities around the world, are there any you don't feel you get along with well or you won't like?" The participants are shown to be fairly frank in their responses. The British man directly states that he is "not a fan of the Germans." The Bangladeshi man answers India and Pakistan due to the historical conflict between the nations. The Kurdish woman answers by stating, "there is a part of me that hates Turkish people…not people but the government." Lastly, an Icelandic man directly tells one of the interviewers that he is "more important than you" and that in his opinion he is "stronger and more important than a lot of people." This and the preceding section are meant to convey that prior to taking a DNA test, these participants had a certain understanding of themselves that they were confident about. Additionally, due to the pride they have in their own identities some seemed to have developed certain prejudices against others, some of which are due to historical conflicts, such as with the Bangladeshi man and the Kurdish woman, and others that stem from nationalist pride.

The participants are then asked if they would like to take a journey based on their DNA. After a second interviewer explains how one inherits DNA from their ancestors (i.e., 50% comes from each parent, and 50% of each of those parent's DNA is inherited from their parents, etc.) the participants are then shown spitting into the Ancestry DNA test tubes. The second interviewer states that "the story of you is in that tube" and asks them what they think the "tube" is going to tell them. They all expect the results to match what they have already told the interviewers, such as that they are 100% Bangladeshi or completely British.

Two weeks later all of the participants are invited back. They are shown entering a large room where they all sit in rows at the back of the room. In front of them is a small table where the two interviewers are seated. The participants are called one at a time to receive the results of their DNA tests. Each participant is shown reading their DNA tests, with many in shock of the results they received as they did not match with how they defined themselves at the beginning. A French woman's results suggested that she has 32% British DNA and the Icelandic man discovered that he had DNA from various populations across Europe. Seemingly coincidental, some of the participants who shared prejudices against certain peoples or nations happened to have those same populations appear in their DNA test. The Kurdish woman had DNA from the caucuses, which the interviewer confirms as referencing Turkey, and the British man had a result that said he had 5% German DNA.

After the revelation of all their DNA test results, the French woman talks directly to the interviewers and says, "I am going to go a bit far right now, but this [DNA testing] should be compulsory. There would be no such thing as extremism in the world if people knew their heritage like that, like who would be stupid enough to think of such a thing as a pure race." As she is talking, we are shown four different women, all of whom are fellow participants. One of the women appears to be Black, two of them appear to be white and one woman is somewhat ambiguous. Closer to the end of the ad, two phrases appear in white font in front of black backgrounds. The first is "You have more in common with the world than you think" and the second "An open world begins with an open mind."

The last phrase signals the goal behind the ad: that DNA testing can give us a different understanding about ourselves and our past that could fight against prejudices and give us an open mind. The ad conveys this message by focusing on the participants who had the most shocking results that conflicted with how they understood themselves before taking the DNA test. The cases of the British man, the Kurdish woman, and the Icelandic man are key examples as they were shown having a strong sense of pride in their own identities that most likely informed the prejudices against other peoples they held. However, it turns out that the same peoples they were prejudiced toward ended up as reference populations in their DNA test results. Therefore, they could be genetically related to the same people they disliked. The message conveyed to the audience is that while you may think you have a solid understanding of yourself and your past; your DNA holds a story about your lineage that might surprise you. In fact, you might find out that some of your ancestors may have been related to people you are prejudiced toward in your own life.

The message is further buttressed by the quote from the French woman in the ad. By saying that there would be no extremism or belief in "pure races" if people knew their heritage, she implies that extremism and prejudice are based on a belief in clear biological divisions between different groups of people. DNA tests, by revealing the story hidden in one's DNA, can show us that our heritages are more complex than we often believe, and we could have ancestors from all over the world; that clear biological divisions among humans do not exist; and no one is 100% anything. Implied in the argument that there would be no extremism if everyone took a DNA test is the idea that one would never be prejudiced against a population with whom they knowingly they could share ancestry. If they realized that in fact one of their ancestors may have been from those populations, and by extension that they are then a descendant of that population, they would no longer be prejudiced against those populations as they would be prejudiced against a part of themselves. It is also implied that racist beliefs persist only because of misinformation or a lack of information and that racist beliefs can be eliminated with the spread

of more accurate information.

This argument relies on treating racism and extremism like xenophobia as simply a

matter of individual prejudicial belief. To say that there would be no such thing as racism if these

DNA tests proved that "pure races" do not exists suggests that racism exists only because of false

beliefs about human biological diversity. It ignores that racism is structural and that expressions

of prejudiced beliefs made by individuals are just one part of a larger social structure that

systemically disadvantages people of color. Additionally, it is not a given that learning about

one's heritage through a DNA test will result in a change in how the test taker sees themselves

and other people. They could disregard the test results or even question the validity of the results.

Panosky and Donovan (2019), in their research on white nationalist forums, have observed that

when faced with genetic ancestry testing information that could challenge their essentialist belief

about whiteness, white nationalists will reject the validity of the test themselves or reinterpret the

results as claiming them as statistical errors or develop historical theories to explain the mixing

of white and European genes with non-white and non-European genes. [130] Therefore, those

committed to extremism could just reinterpret or disregard their test results if they so choose.

They hold on to these beliefs because they lend support for their political goals, not because they

are simply misinformed.

Another ad where the idea that DNA test is shown to fight against racism is the second

half of the Prince Ea ad discussed earlier. The second half of the advertisement touches on

themes similar to the Momondo ad. After rhetorically asking the audience which race box to pick

---

[130] Aaron Panofsky and Joan Donovan, "Genetic Ancestry Testing among White
Nationalists: From Identity Repair to Citizen Science," *Social Studies of Science* 49, no. 5
(2019): 665-669,

based on his DNA test results, Prince Ea goes on to say that "The truth is, who we are is more complex than the labels society gave us." As he says this the word "label" appears three times on the wall behind him. The walls around him then begin to collapse, symbolically showing Prince Ea no longer being confined by a box, to reveal that he is located outside in a location with many trees. He states that "just as we may never judge a book by its cover, we must never judge a person by their skin, because when we use MyHeritage DNA, we find amazing stories hidden within." He goes on to state that humans are a "beautiful tapestry of colors" interconnected with one another as all humans' spring from the roots "of one family tree that goes by the name humanity." As he says this last line, white lines start to rise from the floor behind him. Some of the lines form the outline of trees while the other lines form the outline of double helixes, a symbol associated with DNA. The lines end up merging together to form the word "Humanity" which appears right behind Prince Ea.

Figure 7. Prince Ea in front of the text "Humanity." (MyHeritage DNA, "Humanity.")

Similar to the Momondo ad, this second half of the MyHeritage DNA Ad suggests that DNA tests will give people a better sense of how we are all connected by showing people their ancestral connection with populations across the world that they had no knowledge about prior.

Additionally, it suggests that learning about these connections will then lead people to better understand how all humanity is one big family tree. The power of science and a DNA test will bring us all closer together. This message is represented symbolically near the end of the ad with the image of the white lines representing trees and the white lines representing the DNA double helixes combining to from the word Humanity: the DNA tests (double helix) provide us with more information about our family's ancestors (the family trees) and this in turn will allow us to understand how all of humanity is connected.

Related to this message is the idea that racial categories are simply labels imposed on people by society that limit our understanding of our own pasts and cause us to make judgments about other people. The dilemma posed in the ad is that now that Prince Ea knows about his West African and various European ancestries he does not know which race box to pick as he believes picking one would mean dismissing the others. He resolves the issue by saying that humans are more complex than the labels society puts on us, thereby defining racial categories as simply labels imposed on people by society that are meant to describe a person's ancestry. Describing racial categories in this way erases the fact that racial categories are social and political labels that are the product of the racist structure of a society. He then proceeds to state that we should not judge others by the color of their skin because My Heritage DNA shows us that we each have amazing stories hidden within.

By telling the audience that they should not judge someone "based on their color of their skin" because DNA tests can show us "the amazing stories hidden within," Prince EA seems to be arguing that making assumptions about a person's racial identity is itself a form of discrimination or prejudicial thought because you might be negating the other races that person may belong to. DNA tests, by showing us the stories hidden within, can use the power of genetic testing to move us away from this practice by basing racial identity instead on the mix of genetic ancestry in a person's DNA. In turn, this understanding of racial identity can reveal the fact that more people are "multi-racial" then is often assumed, which could be used as evidence that clear racial divisions along genetic lines do not exist. The ad suggests that proving that these labels are somewhat arbitrary because they do not represent a biological reality could persuade people to see that people have more in common with each other, at least at a genetic level, than different.

Implied within this advertisement as well is the idea that the way we understand racial categories is antiquated. Prince Ea sees having to pick a single racial "box" as restrictive only after he learns about his various ancestries though the MyHertiage DNA test. DNA tests are treated as a new more accurate way to understand the self that is based in science. To view racial categories as outdated labels to describe human diversity ignores that racial categories are the products of a racially structured society. Racial categories persist because systemic advantage for whites and systemic disadvantage for non-whites persists. The understanding of race suggested by the MyHeritage DNA ad neglects this reality and treats race and racial categories as if they begin and end with our understanding of human biology.

The Limitation of DNA Tests "Anti-Racism"

For DNA testing companies to claim that their test can help fight against racism, race must be defined as being mostly categories that describe one's ancestry. When race is understood solely as labels used to describe one's heritage, racism is understood as having only to do with individuals holding prejudiced beliefs about certain heritages that relies on the assumption that there are somewhat clear divisions between different groups of people. DNA tests are then seen as a modern technological innovation that can show us that the divisions along racial, ethnic, or national lines that exist in society do not exist at the genetic level. Our DNA can show us that our ancestry is more complex than we know and that one might be surprised to learn about ancestry they never knew about and to learn that they are genetically related to populations across the globe. As a result, one might find they share ancestry with a population of people against which they might hold prejudices. This revelation might cause them to change their thinking as they realize they may be descendant from the same group they dislike. As more people learn about the genetic links that connect people across the world, the more inclined they will be to ditch racial, ethnic, and national labels and see humanity as one big family tree. Additionally, these ads suggest that DNA test can unlock the hidden history of one's past and that since that knowledge of one's past is rooted in biology and discovered through scientific and technological innovation, it should be seen as a more accurate representation of one's past in comparison to one's family history.

This thinking about race ignores that fact that race is a historically contingent political category, and that prejudice is only one element of the racist structure of a society. However, race and racism are represented as personal identities that describe one's ancestry to support the validity of the DNA test themselves and the claims made in these advertisements. As discussed in chapter one, the main assumption undergirding DNA tests is that there are significant differences in the frequencies of certain genes within different populations across the world. As a result, one is able to use a DNA test to provide estimates of which ancestral populations one is descendent from. However, most genes show up with similar frequencies in various populations, thus making it almost impossible to define certain genes as being representative of specific "ancestral populations." Additionally, the way these companies construct their reference populations often results in the homogenization of large regions with diverse populations, such as with the population labeled "West African" or "Native American" and emphasizes genetic dissimilarity across continents and similarity within continents. The end result is that the test gives the illusion that populations can be genetically defined and that through DNA analysis one is able to learn their "true" origins. Since the test themselves are based on essentialist assumptions, these advertisements reinforce essentialist messages about ethnicity, race, and nationality.

Race is essentialized because while the DNA testing companies claim that they are only providing information about a person's "ethnicity," the ads themselves conflate ethnicity, nationality, and race and interpret DNA test results as racial percentages. Any percentage of African ancestry is interpreted as "Black" ancestry, and any percentage of European ancestry is interpreted as "White" ancestry, and so on. This can result in people making claims that they are a certain percentage "Black" or "White" and potentially even lead to some claiming a multi-racial identity, such as Prince Ea or Livie. Race is then both essentialized and treated as a matter of personal identification since the change in how one identifies racially is based on the results of one's DNA test and not one's social position in a society.

Defining racism as mostly prejudice based on false ideas about human biological diversity allows DNA testing companies to claim that their tests have an important purpose in contemporary society. They can claim that the tests can lead us to a more progressive age where through technological innovation we are able to better understand that humans are more similar to one another than we often believe. Once more people realize that there are no clear biological divisions among humans then they will be less inclined to hold prejudice against others, as the lines between "us" vs. "them" are blurred. DNA companies are able to claim that their tests are important anti-racist tools that can be used to combat the increasing racism and xenophobia being experienced in the world.

These advertisements provide a limited understanding of what racism is and why it persists. Race and ethnicities are treated as labels that mostly work to describe one's ancestry, and racism is boiled down to expressions of prejudiced beliefs held by individuals about people belonging to other races. This erases the fact that the races are the products of racism, specifically racialized societies where the economic, social, and political realms are structured in ways to benefit those marked as white and disadvantage those marked as non-white. What the ads do instead is reify essentialized notions of race and ethnicity, specifically that racial categories emerge from our understandings of human biology and that all that is necessary to claim a certain racial or ethnic category is having some degree of genetic ancestry in one's DNA.

These essentialist ideas not only echo some of the biologically essentialist ideas about race from the height of the eugenics movement, but also end up posing the wrong solution to racism. If racism is seen as the result of individual prejudiced beliefs on the basis of false ideas about human biology, then the solution seems to be to simply educate people about the reality of human biological diversity. The logic follows that once confronted with information that contradicts how they think the world works, such as the fact that there are no "pure" races and that many people have genetic ancestry from many parts of the world, prejudiced individuals will look at the world differently and shed their old beliefs. This understanding of racism relies on the idea that racism emerges simply from people not being able to understand difference. When racism is understood as a part of the structure of a society and a system built deliberately to hoard power and resources for a dominant group, the solution is political struggle. By reinforcing essentialist beliefs and treating race and ethnicity as personal identities divorced from politics, DNA testing advertisements reinforce the system that they are supposedly combating.

CONCLUSION

In this book, I analyzed DNA ancestry testing advertisements to unpack the assumptions about the connections between race, ethnicity, nationality, genetics, and ancestry that shape the claims made by DNA testing companies about their DNA tests. This book contributes to the research on DNA ancestry testing advertisements by exploring two themes not thoroughly explored in previous studies: ancestry travel and DNA tests as anti-racist tools.

I used the introduction chapter to provide a brief overview of the boom in DNA ancestry test sales from the past few years and to outline the theoretical frameworks that informed my analysis of the ads. Specifically, I used Omi and Winant's (2015) theory of racial formation and Mize's (2013) argument that race is a product of racism to build the definitions of race and racism that I relied on one throughout the book. I conceptualized racism as the social structure of a society that distribute resources unequally along racial lines and I conceptualized race as the product of racism that justifies this unequal distribution. Roberts's (2011) *Fatal Invention* and TallBear's (2013) *Native American DNA* served as the main texts that informed how societal assumptions about the meaning of race influence how population geneticists conduct their studies, and the social and political consequences DNA tests can have on communities of color. Additionally, I used Hall's (2012) Encoding/Decoding model as a framework to argue that my analysis offers a potential dominant reading of the messages found in the DNA ancestry testing advertisements.

Before engaging in a textual analysis of the ads, I used the first chapter to outline how histories of racial science and biological essentialism shape the construction of DNA ancestry tests. Synthesizing scholarship from scholars in the fields of biology, anthropology, ethnic studies, and science and technology studies, I provided an overview of the relationship between the biological sciences and racial thinking and how this relationship shapes present day population genetics. While the field of population genetics appeared to offer a departure from the racial science of the early 20th century, in practice many population geneticists stopped using the word "race" but held on to the assumption that humans could be divided into discrete groups based on biological similarity and that racial categories served as helpful approximations of human genetic similarity. These assumptions shape how population geneticists conduct and interpret their studies, such as when population geneticists selectively choose smaller populations within a given continent to stand in for the genetic composition of a whole continent in ways that emphasize difference by excluding populations that live on the border between continents. While the term "race" may not be used, the assumption signified by the concept, that human beings can be separated into distinct clearly defined biologically similar groups, continues to inform population genetics. Since DNA ancestry tests are based on the science of population genetics, the issues present in population genetics are also present in DNA ancestry tests. Therefore, the critiques about population genetics discussed in the chapter informed my analysis of the ancestry testing advertisements in chapters 2 and 3.

Chapter 2 focused on ads that fit within the theme of ancestry travel while chapter 3 focused on the ads that fit within the theme of DNA tests as anti-racist tools. The theories of race outlined in the introduction allowed me to make sense of how race and ethnicity were represented in these advertisements. Using Omi and Winant's theory of racial formation as a framework, the ads can be understood as racial projects that attempt to provide an interpretation of what race and ethnicity mean. In these ads, race and ethnicity are interpreted as mainly being labels that are used to describe a person's ancestral background. The common narrative found in many of the testimonials is that the test taker used to identify racially or ethnically in one way, either because of what their family told them about their ancestry or how others categorized them based on their physical appearance, but then changed how they identify themselves because of the new information they gained about their ancestry through the DNA test. By presenting a change in identity in this way, these ads offer an interpretation of race and ethnicity as being grounded mostly, if not solely, in genetics.

This interpretation does not address how racial and ethnic categories are the products of the social structure of a society that is stratified along racial lines. As both Mize (2013) and Bonilla-Silva (1997) articulate, racial categories are products of a racist society as they are used to justify the unequal distribution of resources. In the ads, the test takers are rarely shown discussing how their identities are tied to their social position in American or European societies. When racism is addressed, it is discussed at the interpersonal level, specifically focusing on how individuals hold prejudiced beliefs and how DNA tests can challenge individuals' beliefs about humanity. By avoiding a structural analysis of race, the ads simultaneously "(re)biologize" race and ethnicity by rooting them in genetics, suggest that racism persists because of people's misguided beliefs about human genetic difference, and promote the idea that the solution to racism is educating people about the genetic similarities we share through DNA ancestry tests.[131]

While these ads have the veneer of being anti-racist, these ads are actually racist racial projects as they keep the racist social structure of society intact. Learning about human genetic difference does not change how goods and resources are distributed or how institutions such as the law, education, and the economy are structured to benefit whites over non-whites. In fact, by rooting race in genetics and not seeing race as a product and necessary tool of racism, the ads

---

[131] Omi, "'Slippin' into Darkness,'" 355.

work to further reinforce the status quo by suggesting that there are innate genetic differences among people of different races.

This echoes the racial science of the early 20th century while appearing more modern and scientifically accurate. However, as Roberts (2011) details in *Fatal Invention*, population genetics and DNA ancestry testing companies, like the racial scientists of the past, are influenced by societal assumptions about the meaning of race and these assumptions inform how they collect, interpret, and represent their data. These ads promote the inaccurate idea that race has meaning at the genetic level which has the consequence of leading people to believe that the racist social structure of a society is product of innate genetic differences. Racial inequality is then seen as unchangeable and/or suggestions for how to address these inequalities are tied back to genetics. The latter can be seen in contemporary debates in the medical field about how race and genetics should be addressed in the medical studies, with some scientists taking it as a given that there is a genetic component to racial categories. A recent example includes a 2021 article from *The New England Journal of Medicine* titled "Embracing Genetic Diversity to Improve Black Health" where the authors state that they believe race "has both a genetic and social component."[132] Therefore, the idea that race exists at the genetic level is already circulating in American society and DNA ancestry testing ads reinforce this idea.

Reducing race and ethnicity to genetics also produces colonialist implications about people in the global south. The idea that DNA tests, which rely on the DNA of currently living people, can offer the test taker information about their genetic ancestry relies on the assumption that contemporary populations around the world, especially indigenous populations, are

---

[132] Akinyemi Oni-Orisan et. al., "Embracing Genetic Diversity to Improve Black Health," *The New England Journal of Medicine* 348, no. 12 (2021):1163,

genetically similar enough to populations from centuries ago that they can serve as proxies for the ancestors of test takers living in the U.S. and Europe. For that to be true, these populations in Africa, Asia, and parts of the Americas would have had to have barely moved or interacted with people from other populations. Contemporary populations are frozen in time and treated as representatives of the past that the American or European traveler comes to visit in order to regain connection to their roots.

Adding to the colonial implications of this practice is the fact that the power to decide who belongs to a given community is taken from these populations and given to the test taker/DNA test. TallBear (2013) illustrates how DNA tests have infringed upon tribal sovereignty by reducing what it means to be an indigenous person in the U.S. to a matter of genetics and treating DNA as a "master molecule" that can stand in for sustained relationships with indigenous peoples.[133] These ads directly impact indigenous communities as they have to respond to claims of belonging to their nations on the basis of a DNA test alone. Sometimes they even have to respond publicly as with the case of the Cherokee Nation responding to Senator Elizabeth Warren's claim to Native ancestry on the basis of a DNA test.[134] Thus, another consequence of the ideas reinforced in these ads is that they promote a power imbalance when it comes to claiming belonging to a community, taking the power from indigenous populations around the world to decide who is a member of their community.

---

[133] TallBear, *Native American DNA*, 70.

[134] Aaron Blake, "Why the Cherokee Nation's Rebuke of Elizabeth Warren Matters," *The Washington Post*, October 16, 2018,

Nash (2002) argues that genealogy can be simultaneously personal and political, that personal reasons for wanting to learn more of one's past often intersects with "wider cultural processes, politics, and social concerns."[135] Consumer interest in DNA ancestry testing in the last few years could be understood as a response to the increasingly socially conscious contemporary moment. In a time where race, gender, sexuality, and class have become commonplace, where social movement organizations like Black Lives Matter are widely recognized, and where terms such as white privilege, intersectionality, and anti-racist have become buzzwords, it would make sense that people would become more conscious about their identities and how they fit within the social structure of a society.

Where do they fit within discussions of privilege and oppression? In what ways can they be better allies? In what ways does their background influence how they navigate the world? These questions might lead people, especially those in privileged positions who might not have had to think about their social position much before, to do some self-reflection and learn more about themselves and their background. The DNA tests could be one way for those who felt disconnected from their past to begin to learn more about who they are. For those who are severed from their past because of colonialism, such as Black Americans, the test could be seen as offering some way of recovering the past and being able to claim a more specific origin. For white Americans whose families have distanced themselves from their European origins over generations, the tests could be a way to develop a more specific ethnic identity that they can claim as opposed to the more nebulous "white" racial identity. For the latter, the claim of an ethnic identity could also serve to address anxieties about discourses around immigration and white racial privilege; if they claim an ethnic identity, they can both say that they are also

---

[135] Nash, "Genealogical Identities," 30.

descendants of immigrants who came from somewhere else and that their ancestors too struggled when they first came to what is today the United States.

This desire to learn more about oneself is also likely fueling the desire for ancestry travel. By learning about their ancestry through the test and by traveling to the places of their ancestors they might hope to feel a more grounded sense of self and of where they came from. For those whose past was severed, the travel could be an attempt to build a connection to their ancestors. However, the idea that the DNA test or the travel will lead to a more solidified sense of self is to some degree idealistic. For some, those ties to the past might be impossible to fully recover. In the case of Black Americans, the test might offer a general idea of the regions but not enough about the specific communities one's ancestors may have been a part of. There is also the possibility that one is not welcomed when they visit the places their ancestors may have been from. Add to this the reality that the test has its own limitations, and it becomes clear that relying on DNA test to learn more about one's past can only go so far. A genetic ancestor cannot stand in for specific ancestors one could name from a family tree.

Limitations and Future Directions

In this book, I offered what I thought were possible dominant readings of select DNA ancestry testing ads. I analyzed the ads by considering how one might interpret the messages in the ads if they take the arguments and assumptions in the ads at face value. Since I did not explore how people interpreted the ads, I did not consider the potential alternative readings that viewers could have of the messages in the ads. Viewers who might engage in a negotiated reading might accept some of the arguments made in the ad while not fully accepting others. I could imagine that a viewer might reject the idea that one should change their racial or ethnic identity because of a DNA test, but they still might believe the test offers an accurate description of a person's true ethnic ancestry. Other viewers might engage in an oppositional reading and re-interpret the messages. Some viewers might completely reject the messages in the ads or find the tests somewhat silly. There could be viewers who develop no interest in the test, do not believe it to be valid, or see it as something recreational that might be fun to do but they do not take very seriously. A study into how people interpret the information in these ads, such as through a survey or a collection of interviews, would shed light on how people are receiving claims made in these ads. Such a study would be able to explore the types of readings viewers make and which type of readings are more common.

Another study could also conduct research on people who have taken DNA tests and explore how they understood their DNA test results and if their interpretation was shaped by the claims made in DNA ancestry testing advertisements. Such a study could explore if test takers have changed how they identified on their basis of their DNA tests, their understandings of the connections between race, ethnicity, nationality, ancestry, and genetics, and their understandings of the root causes of racism and xenophobia. Such as study might reveal to what degree test takers accept the claims made by DNA ancestry testing companies in their marketing.

Another study could explore the experiences of people who have engaged in travel based on their DNA test results. By interviewing those who engaged in this type of travel, the study could explore the travelers' motivation for going on their trip, the expectations they had, and their reflections on their experiences. Such as study might reveal if the way ancestry travel is represented in DNA ancestry ads influences the motivations and expectations of those who engage in ancestry travel.

I want to return to the anecdote of the undergraduate student with the Native relative from the introduction as it serves as an example of various consequences that acceptance of the claims made in DNA ancestry advertisements can have. What strikes me about that moment is the way gene fetishism was put into action. The recognition of someone as a member of an indigenous nation was reduced to whether "Native American DNA" showed up on their test results even though they had a tribal ID card. The fact that indigenous identity is a product of histories of colonialism and social relationships with other indigenous people is erased and replaced with the gene which is seen as the objective arbiter of identity. The fact that the student mentioned that the relative had claimed being indigenous before taking the test and that they had a Tribal ID card near the end of the discussion shows that they treated such information as secondary or insignificant details compared to the results of the DNA test. This signals that DNA tests are seen as more accurate sources of knowledge about a person's ancestral past.

The lack of transparency about what the tests actually test for and the limitations of how the results should be interpreted allows the test to have the veneer of scientific accuracy and objectivity that paints other sources of knowledge, such as family stories, as less reliable and more subjective in comparison. The danger of allowing complex social and political categories to be reduced to being matters of genetics and allowing DNA tests to be seen as the most accurate sources of knowledge about the self is that the social and political categories of race, ethnicity, and nationality are products of relations of power. They are not fixed, and they do not exist in a vacuum. By forgetting this, we run the risk of conceptualizing disparities between dominant and subordinated groups as a consequence of their "genetics" as opposed to the product of the structure of a society that disadvantages some groups for the advantage of others.

While the boom in DNA test sales may have ended, DNA tests continue to shape how race, ethnicity, and nationality are discussed in the U.S. During the time that I wrote this book, I came across many instances, both online and offline, of people discussing their experiences taking DNA tests and how the tests shaped their understanding of their race and ethnicity. I have also seen many cases of people challenging the reliability of DNA tests and how social and political categories like race and ethnicity cannot be reduced to genetics alone. My research has shown that my concerns about the claims made in DNA ancestry tests are warranted as the assumptions in these ads reinforce biologically essentialist ideas about human difference. As DNA tests continue to hold relevance in American society, it is important to remain vigilant about how they reproduce ideas that support dominant ideologies.

www.ingramcontent.com/pod-product-compliance
Lightning Source LLC
LaVergne TN
LVHW041714190726
843493LV00007B/2090